D1482743

Cold and Freezer Storage Manual

other AVI books on refrigeration

other AVI books of interest to food freezers

Cold and Freezer Storage Manual

by **W. R. WOOLRICH**

Dean Emeritus and Professor Emeritus of Mechanical Engineering, The University of Texas

and **E. R. HALLOWELL**

General Manager, Snell Refrigeration Supply

WESTPORT, CONNECTICUT

THE AVI PUBLISHING COMPANY, INC.

1970

Library of Congress Catalog Card Number: 76–115688
1SBN–0–87055–074–8

Printed in the United States of America

Preface

In the more than a century of successful mechanical refrigeration for the cold storage of perishable foods, there has been a "consummation devoutly to be wished" by millions of persons on planet earth. For five thousand years, the forefathers of almost every race on earth have found "God-made" snow and ice useful as a preservative of perishable foods, especially when such cold could be obtained and preserved conveniently for use during the annual hot periods.

Where hot weather seasons prevailed, the total tonnage of foods that could be preserved by cold usually diminished significantly as the summer passed. Only those who lived in cold climates or near massive natural ice warehouses could enjoy year-round cold storage foods. With the advent of mechanical refrigeration, which made continuous cold storage possible, the promise of world wide food preservation facilities became a reality. From the tropical Gulf Coast of the United States to the hot climates in Australia and New Zealand, the new break-through in continuous food preservation by means of mechanical and/or chemical cold became economically feasible.

Prior to World War II, most cold storage warehouses in the United States had a limited demand for freezer storage. Most subfreezing storage was confined to the freezer locker plants. The commercial cold storage business was concerned primarily with the storing of foodstuffs at temperatures above 32°F. In Europe and Great Britain, freezer storage was developed soon after the invention of mechanical cold. This freezer storage was necessary to hold the imported meat shipments that, by virtue of their six-week ocean shipment across tropical equatorial waters, required freezing before loading in the steamship freezer storage holds. Upon the steamships docking in their respective wharves in Europe and Great Britain, the shipments had to be transferred to equally low temperature cold rooms for the period of holding and distribution.

The advent of quick freezing, mostly a post World War I development, changed the ratios of cold and freezer rooms as the demand for freezer space became more acute. While in 1920 many refrigerated warehouses catered to storage above 32°F, today many modern cold storages are designed for freezer storage only at or below 0°F. This great change in the commercial acceptance and need for storage at freezer temperatures has brought a revolution in refrigerated warehouse construction, especially for floors, doors, loading docks, wall structures and insulation.

Over the past few years, the frozen food industry has expanded many fold in freezer storage and has also been penetrated by the kindred fields of locker, underground, chain store and home cabinet freezers. Large plants have continually expanded and very low temperature installations are also increasing.

The following chapters are a manual of storage practice, design criteria and miscellaneous information intended to serve as a guide to anyone interested in cold storage design. It is an attempt to put into lay language various facets of design and operating procedures common to most storage warehouses. It is not intended as a technical manual of exact design but rather as a coordinating volume for use between owner and designer of a cold storage warehouse to effect a better understanding of what may be expected in design and operation.

W. R. Woolrich
E. R. Hallowell

November 14, 1969

Contents

Principles of Refrigeration

CHAPTER 1

Physical Principles of Refrigeration

INTRODUCTION

The dominant problem of ancient man was how to get food, whereas the problem of modern man is how to keep it wholesome after he gets it. Refrigeration presents a procedure of food preservation which is superior to any other method that has ever been known to civilization.

PRINCIPLES OF HEAT

The molecular theory of heat assumes that molecules or particles of substance are in continuous and irregular motion and that heat is the result of this motion. One of the first interests in the study of refrigeration is the transfer of heat. Heat may be transferred in any one of three different ways; or more generally stated, it may be distributed in all three ways at the same time. The three modes of distribution of heat are by conduction, convection, and by radiation.

DEFINITIONS

Evaporation

Evaporation of a pure liquid begins and may continue until all is entirely in the form of vapor. During this period, the temperature of evaporation at saturation remains constant. The heat added during this change of state is the enthalpy of evaporation, made up of the increase in internal energy and the mechanical work done in expanding the liquid to vapor against the constant pressure. The total sum of the enthalpies of saturated liquid and vaporization is the enthalpy of the saturated vapor.

Condensation

Condensation is the reverse of evaporation. It is the change of state of a substance from a gas to a liquid. To prevent condensation on the

1

outside of insulated equipment requires sufficient thickness of insulation to insure that the temperature drop from ambient air to insulated surface in less than the dew point depression. To stop condensation at 100% RH requires an infinite thickness of insulation, thus zero heat flow.

Conduction

By conduction is meant the flow of heat through an unequally heated body or system of bodies, from points of higher to lower temperature. It is exemplified in the heating of a metal rod by placing one end in a flame. The part in the flame soon becomes hot, the molecules of the adjacent parts have their motion quickened through the impact of those in the hotter part, and a transfer of heat takes place to points of lower temperature.

Convection

By convection is meant the transfer of heat by bodily movement of heated particles of matter.

Radiation

By radiation is meant the transfer of energy from point to point in space by means of waves set up in the ether. The earth is heated by radiation from the sun. If one holds his hands over a heated object, the hands are heated both by convection through the air and by radiation, but if the hands are placed under the heated object, the heating of the hands is by radiation only.

Temperature

Temperature may be defined as the thermal condition of a body. Temperature indicates how hot or cold a substance is; that is, it is a measure of sensible heat. However, it does not show how much heat a body may contain. For instance, a bucket containing 5 gal. of water at room temperature 70°F contains more heat than a bucket containing only 1 gal. at 150°F. Temperature, therefore, gives only the intensity of heat, and not the amount. Temperatures are measured by thermometers or pyrometers graduated in Fahrenheit, Centigrade, Rankine or Kelvin scales. The following methods of calculation permit the transfer from Fahrenheit to Centigrade scale or vice versa

Centigrade degrees $= 5/9 \times$ (Fahrenheit degrees — 32 degrees)

Fahrenheit degrees $= (9/5 \times$ Centigrade degrees$) + 32$ degrees

Absolute temperatures are based on absolute zero at which all molecular thermal energy is absent. Numerically, it is 459.69° F below the zero Fahrenheit or 273.16° C below zero Centigrade temperature. In equation relationship these become: $T = t_f + 459.69°$ R for the Fahrenheit scale and $T = t_c + 273.16°$ K for the Centigrade scale.

An ideal way to measure heat is to note its effect in raising the temperature of a measured body of water. The present generally accepted heat unit called the British thermal unit (abbreviated Btu) is defined as 1/180 of the heat required to raise the temperature of 1 lb of water from 32°F to 212°F at normal atmospheric pressure; in other words, in practice, 1 Btu is the measure of that heat which will raise the temperature of 1 lb of water 1°F.

Specific Heat

The specific heat of a substance is the ratio of the heat required to raise the temperature of unit mass of the substance 1° to the heat required to raise the temperature of unit mass of water 1°. American, British and Australian engineers consider the pound as the unit mass of a substance and 1°F as the unit of temperature. Most of the rest of the scientific world measure mass by grams and kilograms and 1°C as the unit of temperature.

FORMS OF HEAT ENERGIES

Sensible Heat

Sensible heat may be defined as that heat which produces a rise of temperature, as when a pan of water placed over a flame becomes hotter and hotter to the touch. One must carefully differentiate this type of heat from another, known as latent heat.

Latent Heat

The quantity of heat required to change the state or condition under which a substance exists, without changing its temperatures; e.g., a definite quantity of heat must be transferred to ice at 32°F to change it into water at the same temperature. This definite quantity of heat is known as the latent heat of fusion in going from the solid to the liquid state, or the latent heat of evaporation when going from the liquid to the vapor state, as when water boils and forms steam.

It will readily be seen that latent heat is of great importance in the study and application of refrigeration. Water, when cooled, loses only 1 Btu per lb for each degree decrease in temperature from whatever temperature it is, until it reaches 32°F. Then 144 Btu (which is the latent heat of fusion for water) are extracted while the water is freezing, yet there is no change in the temperature of the water during this period. When the water is all frozen, the resulting ice then requires only approximately one-half Btu per lb for each degree decrease in temperature below 32°F. The process is, of course, reversible, and therefore, to change 1 lb of ice to water, it is necessary to absorb 144 Btu. Naturally then, it is the latent heat of fusion of ice that make it valuable for refrigeration.

The latent heat of evaporation is even more important in the study of and application of refrigeration, for without this phenomenon it would be impossible to have mechanical refrigeration by the compression system. It is the latent heat of certain substances known as refrigerants, that forms the basis of producing refrigeration by mechanical means.

Although thus far no real definition has been given for refrigeration, some points have been listed that do define it. In the literature numerous definitions are given, and a combination of a number of these would indicate that it is a process of removing heat from a confined space and material for the purpose of reducing and maintaining the temperature below that of the surrounding condition. Since refrigeration is a process whereby heat is removed, the specific quantity of heat removed is measured in Btu.

REFRIGERATION UNITS

The standard unit of refrigerating capacity is known as a ton of refrigeration. The ton of refrigeration is derived on the basis of the removal of the latent heat of fusion from 2,000 lb of water at 32°F in order to produce 2,000 lb of ice at the same temperature in 24 hr. The latent heat of fusion of ice (by calorimeter 143.4) is accepted as being 144 Btu per lb. Therefore, with 2,000 lb of water at 32° F and the extraction of 144 Btu from each pound, a total of 288,000 Btu are removed to change a ton of water to a ton of ice at 32°F. The standard ton of refrigeration is therefore: 288,000 Btu per 24 hr, or 12,000 Btu per hr, or 200 Btu per min.

THE COMPRESSION CYCLE OF REFRIGERATION

In the compression type of the mechanical refrigeration process the gaseous refrigerant is compressed then passed through pipes to the condenser where it is cooled and condensed to a liquid minus much of its original heat. Usually the liquid refrigerant is stored in a high pressure cylindrical receiver with inlet and outlet valves in the continuous piping system.

Under flow rate control the liquid refrigerant at high pressure is passed from the receiver through an expansion throttling pressure reducing valve. The resultant action is to change the fluid to an atomized vapor-liquid mixture at low pressure as it enters the evaporator or cooler.

Actually the evaporator or cooler is a boiler in which the vapor-liquid refrigerant is completely vaporized by heat obtained from the enclosed refrigerator product and produce, which in turn warms the fluid refrigerant in the cooler coils to a gaseous condition. When the cycle has been completed by the refrigerant in the closed refrigeration system, the gaseous vapor starts again over the same path changing by compression, cooling and heating from a gas to a liquid and returning again as a gas to the compressor for this and subsequent cycles.

STANDARD TON OF REFRIGERATION

The standard rating of a refrigerating system using liquefiable gas or vapor is the number of standard tons of refrigeration it performs under adopted pressures of refrigerants, namely; the inlet (suction) pressure being that which corresponds to a saturation temperature of 5°F and the discharge pressure being that which corresponds to a saturation temperature of 86°F. In the case of ammonia, this would correspond to a suction pressure of 20 lb gage and a discharge pressure of 155 lb gage.

COEFFICIENT OF PERFORMANCE

The coefficient of performance of a refrigerant in the closed cycle denotes a measure of the efficiency of operation in utilizing the energy input. This is the ratio of the energy utilized in the evaporator to the energy input of compression.

COMPRESSION RATIO

Under normal conditions, the refrigerant with the lowest compression ratio is preferable, since differences between the suction and discharge temperature reflect the boiling point characteristic of the refrigerant adopted, and further, since the temperature of the evaporator and condenser are prescribed by system design and the available condensing fluid heat level, then compression ratio is determined by the refrigerant selected.

The compression ratio of a refrigerant compressor is the quotient of the initial to the final volume within the cylinder when the piston is at its maximum discharge position. Because volume varies inversely as the pressure under constant temperature conditions the compression ratio may be defined as the final pressure divided by the initial pressure.

In the first instance, the equation becomes: Compression ratio = initial volume/final volume. Under the second condition the equation becomes: Compression ratio = final pressure/initial pressure.

The latter equation is readily applicable to the computing of the compression ratios of rotary and centrifugal machines. Both equations are readily applicable to all reciprocating compressors.

HEAT TRANSFER IN PIPES FOR EVAPORATOR AND CONDENSER DESIGNS

In the transfer of liquid or gas within a metal pipe to a liquid or gas or through the solid metal wall, the factors affecting the overall heat movement are numerous. In the case of new metal pipe carrying a fluid on the inside and another fluid moving over the outside, two different films affecting the heat transfer are present.

In the most simple system, there exists, (a) heat moving within the pipe by convection, (b) heat moving through the inside film by convection, (c) heat moving through the metallic pipe wall by convection, (d) heat moving through the film on the outside by conduction, then (e) heat moving beyond the film on the outside by convection.

Gas films of equal thickness offer more resistance than liquid films, but usually liquid films are of greater thickness than gas.

The overall U value of condensers and evaporators is a summation of all of these factors expressed in Btu per hr per sq ft per °F temperature difference.

Specification Design Values for Evaporators

Well-designed evaporators should possess (a) high heat transfer rate, and (b) unobstructed free flow of refrigerant and coolant.

The refrigerant side depends on complete wetting of the metal surfaces and the ease with which the gas formed as the refrigerant boils and then can escape freely to the suction lines. It should "boil off" easily.

Improved capacity is assured when small tubes are employed, maximum turbulence is maintained, the differential temperatures are high and all tubes are submerged and clean of oil and scale.

HEAT OF LOADED ELECTRIC MOTORS WITHIN REFRIGERATED SPACE

One mechanical horsepower equals 2,545 Btu of heat produced per hour. One kilowatt of electrical power produces 3,413 Btu of heat per hour. Allowing for the electric losses in driving a compressor fan: A 3 to 10 hp motor fully loaded will give off 3,000 Btu per hp hr. This will require 1 ton of refrigeration to be allowed for each 4 hp of motor load.

A 1 to 3 hp fully loaded motor will give off approximately 3,500 Btu per hp hr. For estimating, allow 3.3 hp per hr per ton of refrigeration.

TABLE 1.1

SOME AVERAGE OVERALL HEAT TRANSFER COEFFICIENT U VALUES FOR CONDENSERS

	U
Ammonia condenser—atmospheric counter flow type	150
Ammonia condenser—double pipe type	250
Ammonia condenser—multi-pass type	250
Ammonia condenser—shell and tube—vertical	150
Double pipe heat exchanger	80

TON OF REFRIGERATION FOR WELL-INSULATED COLD STORAGE, FREEZER STORAGE AND SHARP FREEZER STORAGE PLANTS

For small storage plants of 10,000 cu ft capacity, allow 4 tons for cold storage and 8 tons for sharp freezer storage plants.

For storage plants of 40,000 cu ft capacity, allow 8 tons for cold storage, and 20 tons for sharp freezer storage.

For storage plants of 100,000 cu ft capacity allow 14 tons of refrigeration for cold storage, and 25 tons for sharp freezer storage.

TABLE 1.2

COMPRESSION RATIOS OF COLD STORAGE REFRIGERANTS
STANDARD TON CONDITIONS

Refrigerant	Gage Pressure, psi		Compression Ratio
	At 86 °F Discharge	At 5 °F Suction	
Carbon dioxide	1,024.3	319.7	3.11
R. 22	159.8	28.33	4.05
Ammonia	154.5	19.57	4.94
R. 12	93.2	11.81	4.07
Methyl chloride	80.83	6.19	4.57
R. 114	21.99	16.14[1]	5.42
R. 21	16.53	19.25[1]	5.98
R 11	3.58	23.95[1]	6.20
Methylene chloride	9.44[1]	27.53[1]	8.57
R. 113	13.93[1]	27.92[1]	8.01

[1] Inches of mercury below atmospheric pressure.

TABLE 1.3

COMPARISON OF PHYSICAL THERMAL VALUES OF MOLECULAR WEIGHTS OF MOST COMMONLY USED REFRIGERANTS FOR COLD WAREHOUSING

R No.	Name	Molecular Weight	Freezing Point °F	Boiling Point at 1 Atm °F	Critical Temperature °F	Critical Pressure Psia	Underwriters Higher Safety Ratings
40	Methyl chloride	50.48	−143.7	−10.76	421	640	4
717	Ammonia	17.03	−107.9	−28.0	271.2	1651	2
11	Trichloromonofluoromethane	137.4	−168	74.7	388.4	635	5
12	Dichlorodifluoromethane	120.9	−252.4	21.6	2327	582	6
13-B-1	Monobromotrifluoromethane	148.9	−270	72	152.6	575	5
21	Dichloromonofluoromethane	102.93	−211	48	353.3	750	b4
22	Monochlorodifluoromethane	86.48	−256	−41.4	204.8	716	5
113	Trichlorotrifluoroethane	187.39	−31.0	117.6	417.4	405	b4
114	Dichlorotetrafluoroethane	170.93	−1.37	38.4	294.3	474	6
500	12/152a Azeotrope	99.29	−254	−28	221	631	
502	12/115 Azeotrope	111.64		−50	194	618	
744	Carbon dioxide	44.0	−69.9	−108.4	87.8	1071	5

TABLE 1.4

PHYSICAL CHARACTERISTICS OF THE MOST COMMONLY USED REFRIGERANTS UNDER AMERICAN STANDARD TON CONDITIONS
(5°F SUCTION—86°F CONDENSATION)[1]

Refrigerant No.	Name	Suction Pressure Psig	Condensing Pressure Psig	Compression Ratio	Net Refrigerating Effect Btu/Lb	Refrigerant Circulated Lb Min	Compressor Displacement Cfm	Horsepower Hp per Ton	Coefficient of Performance	Comp. Discharge Temp. °F
40	Methyl chloride	6.5	80.0	4.48	150.2	1.33	5.95	0.962	4.90	172
717	Ammonia	19.6	154.5	4.94	474.4	0.422	3.44	0.989	4.76	210
11	Trichloromonofluoro-methane	24.0*	3.6	6.24	67.3	2.98	36.46	0.933	5.05	109
12	Dichlorodifluoro-methane	11.8	93.3	4.08	50.0	4.00	5.83	1.002	4.70	101
13-B-1	Monobromotrifluoro-n ethane	63.2	247.1	3.36	29.3	6.86	2.63	1.030	4.25	124
21	Dichloromonofluoro-methane	19.2*	16.5	5.96	89.4	2.24	20.43	0.941	5.01	142
22	Monochlorodifluoro-methane	28.2	158.2	4.03	70.0	2.86	3.55	1.011	4.66	128
113	Trichlorotrifluoro-ethane	27.9*	13.9*	8.02	53.7	3.73	102.3	0.973	4.84	86
114	Dichlorotetrafluoro-ethane	16.1*	22.0	5.42	43.1	4.64	20.14	1.049	4.49	86
500	12/152a Azeotrope	16.4	113.4	4.12	61.1	3.27	4.97	1.022	4.61	105
502	22/115 Azeotrope	36.0	175.1	3.75	45.7	4.38	3.61	1.079	4.37	99
744	Carbon dioxide	317.5	1031	3.15	55.5	3.62	0.96	1.840	2.55	151

[1] Saturated suction vapor for Refrigerants 113, 114. In these cases, enough suction superheat was assumed to give saturated discharge vapor.
* Inches of mercury vacuum.

CHAPTER 2

Physical Characteristics of Commercial and Cryogenic Refrigerants and of Atmospheric Air for Cold and Freezer Rooms

BOILING AND CONDENSING TEMPERATURES AND PRESSURES

The physical, chemical and thermodynamic properties of the several hundred available refrigerants determine their practical usefulness. The more important of the physical characteristics will be considered first.

In dealing with refrigerant fluids, the evaporator and condensing temperatures determine the pressures. For most applications, it is desirable to select a refrigerant whose saturation pressure at minimum evaporating operating temperature is maintained at a pressure a few pounds above atmospheric. This gives a positive differential between suction and atmospheric pressure, and prevents leakage of air with inherent moisture into the low pressure side of the system, especially at the compressor shaft and piston rod seals. The maximum condensing temperature is largely affected by climatic conditions. It is preferable to adopt a refrigerant with a condensing pressure within safety limitations and within system weight acceptability. Usually an air cooled condensing system will inherently require higher condensing temperatures especially in hot climates. High condensing pressures are conducive to more system leakage and accidents.

FREEZING, CRITICAL AND DISCHARGE TEMPERATURES OF REFRIGERANTS

The refrigerant fluid should have a low freezing temperature to avoid operational obstruction by the refrigerant itself. At the other end of the scale it is most desirable that the critical temperature be well above the maximum condensing temperature. Exceptions to refrigerants not having critical temperatures above the temperature of the condenser are air and carbon dioxide, the latter when the condensing medium is above 87.8°F. Under these temperatures machines operate as "dry gas" systems at a much lower efficiency.

High discharge temperatures from the compressor are the cause of some refrigerant breakdowns and poor lubrication effectiveness and should be avoided whenever possible.

TABLE 2.1

FOR STANDARD TON CONDITIONS PRESSURES AND QUANTITIES
OF REQUIRED COLD STORAGE REFRIGERANTS

Refrigerant	Boiling Point, °F	Lb Liquid Refrigerant Evaporated per Min	Vapor Volume per Lb 5 °F	Cu Ft Displacement per Min
Carbon dioxide	−108.4	3.528	0.2673	0.943
R. 22	− 41.4	2.887	1.246	3.596
Ammonia	−28.0	0.4215	8.150	3.436
R. 12	− 21.6	3.916	1.485	5.815
Methyl chloride	− 10.76	1.331	4.471	5.95
R. 114	38.4	4.640	4.221	19.587
R. 21	48.0	2.237	9.132	20.427
R. 11	74.7	2.961	12.27	36.33
Methylene chloride	103.7	1.492	49.9	74.45
R. 113	117.6	3.726	27.04	100.76

LATENT HEAT OF VAPORIZATION AND SPECIFIC HEAT OF THE REFRIGERANT

Since most refrigerants, as used, pass through a liquid and vapor cycle from receiver to compressor and heat absorbed by them per pound is mostly the heat of vaporization, the higher the latent heat capacity of the refrigerant, the less gas must be compressed. Other factors that must be given consideration are the specific heat of the liquid and the densities of both the refrigerant liquid and the vapor. For small units, the latent heat capacity of the refrigerant is less a favorable factor than for large compressor operation.

TABLE 2.2

OPERATING PHYSICAL CHARACTERISTICS OF PREFERRED COLD STORAGE REFRIGERANTS

	Heat Content Btu per Lb		Btu Refrigerating Effect per Lb	Lb/Min Liquid Refrigerant Circulated per Ton 86 °F 5 °F	Cu In. Liquid Refrigerant To Be Evaporated per Min
	Vapor 5 °F	Liquid 86 °F			
Ammonia	613.35	138.9	474.45	0.4215	0.02691
Methylene chloride	163.8	29.75	134.05	1.492	0.01198
Methyl chloride	196.92	46.67	150.25	1.331	0.01778
R. 21	119.97	30.56	89.41	2.237	0.01183
R. 11	92.88	25.34	67.54	2.961	0.01094
R. 113	79.60	25.93	53.67	3.726	0.01031
R. 22	105.56	36.28	69.28	2.887	0.01363
R. 12	78.79	27.72	51.07	3.916	0.0124
R. 114	72.21	29.11	43.10	4.640	0.0112
Carbon dioxide	102.14	45.45	56.69	3.528	0.0267

TABLE 2.3

FREEZING POINTS OF PREFERRED COLD STORAGE REFRIGERANTS

Refrigerant	Freezing Point ,°F
R. 22	−256
R. 12	−252.4
R. 21	−211
R. 11	−168
Methyl chloride	−143.7
Methylene chloride	−143
R. 114	−137
Ammonia	−107.9
Carbon dioxide	− 69.9
R. 113	− 31.0

High specific heat of the refrigerant vapor is a desirable factor. The heat that is added to the saturated vapor in the evaporation and suction lines will boost its temperature less if the specific heat is low. A high specific heat of the refrigerant vapor is an asset in practical operation. On the other hand, a low specific heat of the refrigerant liquid is an asset since less latent heat of evaporation in passing through the expansion valve must be used in cooling the liquid. This leaves more cooling capacity in the evaporator.

COMPRESSION RATIO AND THE REFRIGERANT

A low compression ratio is recommended to hold down first cost of the compressor and to reduce the operation energy required. The compression ratio as previously defined, affects design, construction and operating characteristics of the entire system. Rotary and centrifugal

TABLE 2.4

REFRIGERATING EFFECT AND QUANTITY OF REFRIGERANT REQUIRED PER TON
(STANDARD TON CONDITIONS)

Refrigerant	Refrigerating Effect Btu per Lb (Standard Cycle 5 °F to 86 °F)	Weight Refrigerant Circulated per Standard Ton, Lb per Min
Ammonia	474.5	0.422
Methyl chloride	150.3	1.331
Sulfur dioxide	141.4	1.414
Dichloromonofluoromethane (R. 21)	89.4	2.237
Monochlorodifluoromethane (R. 22)	69.3	2.887
Trichloromonofluoromethane (R. 11, Carrene-2)	67.5	2.961
Carbon dioxide	56.7	3.528
Trichlorotrifluoroethane (R. 113)	53.7	3.726
Dichlorodifluoromethane (R. 12)	51.1	3.916

refrigerant compressors operate better on fluids having low ratios of compression, especially to reduce vapor leakage by their rotors.

LIQUID DENSITIES AND VISCOSITIES OF REFRIGERANTS

Density must be considered with viscosities in refrigerant compressor operation. Most designers prefer low vapor density refrigerants in order to justify high gas velocities in the suction and discharge lines and valves. Liquid densities affect the operation of float valves and some refrigerant oil mixtures. The trend is towards higher liquid densities in refrigerant use. For long lines, low viscosity of the liquid refrigerant is desired to reduce pressure drop in the orifices and lines. Likewise to reduce pressure drop and line size low vapor viscosities are usually preferable.

CHEMICAL CHARACTERISTICS OF REFRIGERANTS

The chemical characteristics of refrigerants are most important, not only for thermodynamic considerations, but for fire, explosion, safety and odor considerations. These are especially important in freezer storage on land and on marine craft.

TOXICITY OF REFRIGERANTS

The toxicity of refrigerants is rated by Fire Underwriters Laboratories. This rating is based on the toxic effect on humans over specified periods. Carbon dioxide, air, nitrogen, nitrous oxide and fluorocarbon refrigerants have preferred ratings because of very low toxicity at normal refrigerating temperatures. Carbon dioxide, Refrigerants 12, 22 and 502 have most acceptable ratings for marine service on account of their preferred underwriter's ratings as nontoxic, nonirritating properties on shipboard.

FLAMMABILITY AND EXPLOSION HAZARD

The flammability and explosive hazard of many potential refrigerants are cause for rejection of many fluids of excellent thermodynamic refrigerating properties. The particular hazard of consequence is the possibility of leaks occurring in the refrigeration system, bringing about explosive concentrations of the flammable vapor with air.

Ammonia and methyl chloride will burn but are explosive only under unusual conditions.

Carbon dioxide, nitrogen, nitrous oxide and the fluorocarbon refrigerants are nonflammable and nonexplosive.

Detection of the location of refrigerant leaks is a chemical reaction search. Fluorocarbon leaks can be traced with a halide torch. Methyl chloride leaks can be traced by adding 1% acrolein to the refrigerant, then detection is by the escaping odor. Ammonia leaks are readily noted by their odor and final location is improved by burning a sulfur candle and noting the ammonium sulfite smoke, at the leak location. Sulfur dioxide leaks are most readily detected by sponges or cloths soaked in aqua ammonia (25%).

REFRIGERANT ODORS

Odors of refrigerants can be both an asset and a hazard. Odors of a refrigerant make it easy to detect leaks. But the same odors may contaminate foodstuffs in storage at all temperatures.

The fluorocarbons, carbon dioxide, nitrous oxide, air and nitrogen rank very high as nearly odorless refrigerants. Ammonia is a compound that is not only toxic to human beings, but must be carefully isolated because very small percentages give an odor to foods and stored products.

SATURATED REFRIGERANTS

Explanation of Terms Used in Tables of Properties of Saturated Refrigerant

(Note that these tables are based on a temperature of −40°F)

(1) **Temperature in Degrees Fahrenheit.**—Given in the first column is the temperature of the saturated ammonia. The properties tabulated are for saturated ammonia, and will not hold for superheated conditions. Saturated ammonia contains no liquid ammonia and has no superheat.

(2) **Absolute Pressure in Pounds per Square Inch.**—Absolute pressure is measured from zero. If a perfect vacuum existed, the absolute pressure would be zero.

(3) **Gage Pressure in Pounds per Square Inch.**—Gage pressure is measured from 14.7 lb abs or, in other words, it begins at atmospheric pressure as zero.

(4) **Specific Volume of the Liquid in Cubic Feet.**—This gives the number of cubic feet of ammonia liquid in 1 lb at this temperature.

(5) **Density of Vapor.**—The density of a vapor is found by dividing *one* by the number of cubic feet of the vapor in a pound at that pressure.

For example, the density given in the tables for a temperature of 10°F as 0.1369 is found by dividing 1 by 7.304, the volume given in column 4.

(6) **Enthalpy or Heat Content of the Liquid.**—This column gives the total heat in the liquid ammonia, assuming —40°F as the zero from which the heat is computed.

(7) **Enthalpy or Heat Content of Vapor.**—This column gives the total heat of the vapor above —40°F. These values are found by adding the heat content of liquid in column 6 to the latent heat of vaporization of ammonia in column 8.

(8) **Latent Heat of Vaporization of Ammonia.**—This column gives the amount of heat required to vaporize the ammonia after it has been raised to its boiling point at this pressure and temperature.

(9) **Entropy of the Liquid.**—These values are based on the total heat of the liquid found in column 6. These are computed from above —40°F.

(10) **The Entropy of the Vapor.**—This includes the entropy of liquid and latent heat based on the temperature above —40°F.

Problem 1. What volume will 10 lb of ammonia occupy at 100°F? Computation. Referring to Table 2.5, the volume of 1 lb of ammonia vapor at 100°F is 1.419 cu ft. Ten pounds will, therefore, occupy 10 times 1.419, or 14.19 cu ft.

Problem 2. How much heat must be extracted to lower 20 lb of ammonia vapor from 80° to 0°F, assuming the pressure is likewise lowered? Computation. Referring to Table 2.5, the total heat of 1 lb of vapor at 80°F is 630.7 Btu. The total heat at 0°F is 611.8 Btu. The amount of heat it is necessary to remove from 1 lb will be, then 630.7 —611.8 or 18.9 Btu. For 20 lb it will be 20 × 18.9, or 378 Btu.

THE CRYOGENS AS REFRIGERANTS

The cryogens may be considered the very low temperature refrigerants. Many of the cryogens used today were considered permanent gases until the nineteenth century. By very low temperature research each cryogen was found to have a temperature at which it would liquefy. Above this temperature it could not exist as a liquid. This is known as its critical temperature. Below this critical temperature each cryogen has a boiling and a freezing point not unlike the more common liquids like water.

In 1823, Michael Faraday liquefied carbon dioxide and chlorine. These early experimenters could not liquefy oxygen, nitrogen and hydrogen even when they subjected them to several thousand atmospheres. They called them permanent gases. Critical temperature and pressure were not understood at that date. The critical temperature was not under-

TABLE 2.5

PROPERTIES OF SATURATED AMMONIA

(COMPUTED AND ARRANGED FROM THE U. S. BUREAU OF STANDARDS BULLETIN)

Temp °F t	Pressure		Volume Vapor Ft³/Lb V	Density Vapor Ft³/Lb 1V	Enthalpy From −40°			Entropy		Temp °F
	Absolute Lb/In² p	Gage Lb/In² gp			Liquid Btu/Lb h	Vapor Btu/Lb H	Latent Heat Btu/Lb L	Liquid Btu/Lb°F s	Vapor Btu/Lb°F S	
−60	5.55	¹18.6	44.73	0.02235	−21.2	589.6	610.8	−0.0517	1.4769	−60
−55	6.54	¹16.6	38.38	0.02605	−15.9	591.6	607.5	−0.0386	1.4631	−55
−50	7.67	¹14.3	33.08	0.03023	−10.6	593.7	604.3	−0.0256	1.4497	−50
−45	8.95	¹11.7	28.62	0.03494	−5.3	595.6	600.9	−0.0127	1.4368	−45
−40	10.41	¹8.7	24.86	0.04022	0.0	597.6	597.6	0.0000	1.4242	−40
−35	12.05	¹5.4	21.68	0.04613	5.3	599.5	594.2	0.0126	1.4120	−35
−30	13.90	¹1.6	18.97	0.05271	10.7	601.4	590.7	0.0250	1.4001	−30
−25	15.98	1.3	16.66	0.06003	16.0	603.2	587.2	0.0374	1.3886	−25
−20	18.30	3.6	14.68	0.06813	21.4	605.0	583.6	0.0497	1.3774	−20
−15	20.88	6.2	12.97	0.07709	26.7	606.7	580.0	0.0618	1.3664	−15
−10	23.74	9.0	11.50	0.08695	32.1	608.5	576.4	0.0738	1.3558	−10
−5	26.92	12.2	10.23	0.09780	37.5	610.1	572.6	0.0857	1.3454	−5
0	30.42	15.7	9.116	0.1097	42.9	611.8	568.9	0.0975	1.3352	0
5	34.27	19.6	8.150	0.1227	48.3	613.3	565.0	0.1092	1.3253	5
10	38.51	23.8	7.304	0.1369	53.8	614.9	561.1	0.1208	1.3157	10
15	43.14	28.4	6.562	0.1524	59.2	616.3	557.1	0.1323	1.3062	15
20	48.21	33.5	5.910	0.1692	64.7	617.8	553.1	0.1437	1.2969	20
25	53.73	39.0	5.334	0.1875	70.2	619.1	548.9	0.1551	1.2879	25
30	59.74	45.0	4.825	0.2073	75.7	620.5	544.8	0.1663	1.2790	30
35	66.26	51.6	4.373	0.2287	81.2	621.7	540.5	0.1775	1.2704	35
40	73.32	58.6	3.971	0.2518	86.8	623.0	536.2	0.1885	1.2618	40
45	80.96	66.3	3.614	0.2767	92.3	624.1	531.8	0.1996	1.2535	45
50	89.19	74.5	3.294	0.3036	97.9	625.2	527.3	0.2105	1.2453	50
55	98.06	83.4	3.008	0.3325	103.5	626.3	522.8	0.2214	1.2373	55
60	107.6	92.9	2.751	0.3635	109.2	627.3	518.1	0.2322	1.2294	60
65	117.8	103.1	2.520	0.3968	114.8	628.2	513.4	0.2430	1.2216	65
70	128.8	114.1	2.312	0.4325	120.5	629.1	508.6	0.2537	1.2140	70
75	140.5	125.8	2.125	0.4707	126.2	629.9	503.7	0.2643	1.2065	75
80	153.0	138.3	1.955	0.5115	132.0	630.7	498.7	0.2749	1.1991	80

85	166.4	151.7	0.5552	137.8	631.4	493.6	0.2854	1.1918	85
90	180.6	165.0	0.6019	143.5	632.0	488.5	0.2958	1.1846	90
95	195.8	181.1	0.6517	149.4	632.6	483.2	0.3062	1.1775	95
100	211.9	197.2	0.7048	155.2	633.0	477.8	0.3166	1.1705	100
105	228.9	214.2	0.7615	161.1	633.4	472.3	0.3269	1.1635	105
110	247.0	232.3	0.8219	167.0	633.7	466.7	0.3372	1.1566	110
115	266.2	251.5	0.8862	173.0	633.9	460.9	0.3474	1.1497	115
120	286.4	271.7	0.9549	179.0	634.0	455.0	0.3576	1.1427	120
125	307.8	293.1	1.028	185.1	634.0	448.9	0.3679	1.1358	125

¹ Inches of mercury below one standard atmosphere (29.92 in.).

TABLE 2.6

PROPERTIES OF CHLOROFLUOROMETHANES AND ETHANES COMMONLY USED AS REFRIGERANTS IN COLD AND FREEZER WAREHOUSES

	R. 11 Trichloro-monofluoro-methane	R. 12 Dichloro-difluoro-methane	R. 21 Dichloro-monofluoro-methane	R. 22 Monochloro-difluoro-methane	R. 113 Trichloro-trifluoro-ethane	R. 114 Dichloro-tetrafluoro-ethane
Chemical formula	CCl_3F	CCl_2F_2	$CHCl_2F$	$CHClF_2$	$C_2Cl_3F_3$	$C_2Cl_2F_4$
Freezing point, °F	−168	−252	−211	−256	−31	−137
Boiling point at 1 atm, °F	74.7	−21.6	48.0	−41.4	117.6	38.4
Conditions at critical point:						
Temperature, °F	388.4	232.7	353.3	204.8	417.4	294.3
Pressure, psia	635	582	750	716	495	474
Specific heat of vapor at constant pressure at 1 atm, °F	0.1374 at 110	0.1446 at 68	0.1364 at 60	0.1499 at 68	0.1633 at 160	0.1629 at 110
Specific heat of vapor at constant volume at 1 atm, °F	0.1213 at 110	0.1273 at 68	0.1156 at 60	0.1261 at 68	0.1516 at 160	0.1502 at 110
Latent °F heat of vaporization at 5°F	84.00	69.47	109.34	93.45	70.62	61.98
Suction pressure at 5°F, psia	2.931	26.51	5.243	48.12	0.9802	6.77
Condensing pressure at 86°F, psia	18.28	107.9	31.23	174.2	7.856	36.69
Coefficient of performance, standard rating cycle	4.93	4.72	4.89	4.68	4.93	4.75
Toxicity, Underwriters Laboratories Grouping	5	6	4	5A	6	6

TABLE 2.7

PROPERTIES OF SATURATED VAPOR

OF DICHLORODIFLUOROMETHANE (REFRIGERANT 12)

USED EXTENSIVELY IN COLD AND FREEZER WAREHOUSING

Temp °F t	Pressure		Volume		Density		Enthalpy			Entropy		Temp °F t
	Absolute Psi p	Gage Psi gp	Liquid Ft³/Lb vf	Vapor Ft³/Lb vg	Liquid Lb/Ft³ 1/vf	Vapor Ft³/Lb 1/vg	Liquid Btu/Lb f	Latent Btu/Lb hfg	Vapor Btu/Lb hg	Liquid Btu/Lb°F sf	Vapor Btu/Lb°F sg	
−155	0.1163	29.68[1]	0.00954	232.29	104.86	0.004305	−24.61	84.61	60.00	−0.0686	0.2092	−155
−150	0.1527	29.61[1]	0.00957	179.79	104.46	0.005562	−23.50	84.07	60.57	−0.0650	0.2065	−150
−140	0.2554	29.40[1]	0.00965	110.92	103.64	0.009016	−21.29	83.01	61.72	−0.0580	0.2017	−140
−130	0.4116	29.08[1]	0.00973	70.94	102.80	0.01410	−19.10	81.98	62.88	−0.0512	0.1975	−130
−120	0.6417	28.61[1]	0.00981	46.84	101.95	0.02135	−16.94	80.98	64.04	−0.0448	0.1937	−120
−110	0.9709	27.94[1]	0.00989	31.84	101.08	0.03141	−14.78	80.00	65.22	−0.0385	0.1903	−110
−100	1.430	27.01[1]	0.00998	22.20	100.20	0.04504	−12.64	79.04	66.40	−0.0325	0.1873	−100
−90	2.054	25.74[1]	0.01007	15.86	99.30	0.06305	−10.51	78.10	67.59	−0.0266	0.1847	−90
−80	2.885	24.05[1]	0.01016	11.57	98.39	0.08640	−8.40	77.17	68.77	−0.0210	0.1823	−80
−70	3.971	21.84[1]	0.01026	8.608	97.46	0.1162	−6.30	76.25	69.95	−0.0155	0.1802	−70
−60	5.365	19.00[1]	0.01036	6.516	96.51	0.1535	−4.20	75.33	71.13	−0.0102	0.1783	−60
−50	7.125	15.42[1]	0.01047	5.012	95.55	0.1995	−2.11	74.42	72.31	−0.0050	0.1767	−50
−40	9.32	10.92[1]	0.0106	3.911	94.58	0.2557	0	73.50	73.50	0	0.17517	−40
−30	12.02	5.45[1]	0.0107	3.088	93.59	0.3238	2.03	72.67	74.70	0.00471	0.17387	−30
−20	15.28	0.58	0.0108	2.474	92.58	0.4042	4.07	71.80	75.87	0.00940	0.17275	−20
−10	19.20	4.50	0.0109	2.003	91.57	0.4993	6.14	70.91	77.05	0.01403	0.17175	−10
0	23.87	9.17	0.0110	1.637	90.52	0.6109	8.25	69.96	78.21	0.01869	0.17091	0
10	29.35	14.65	0.0112	1.351	89.45	0.7402	10.39	68.97	79.36	0.02328	0.17015	10
20	35.75	21.05	0.0113	1.121	88.37	0.8921	12.55	67.94	80.49	0.02783	0.16949	20
30	43.16	28.46	0.0115	0.939	87.24	1.065	14.76	66.85	81.61	0.03233	0.16887	30
40	51.68	36.98	0.0116	0.792	86.10	1.263	17.00	65.71	82.71	0.03680	0.16833	40
50	61.39	46.69	0.0118	0.673	84.94	1.485	19.27	64.51	83.78	0.04126	0.16785	50
60	72.41	57.71	0.0119	0.575	83.78	1.740	21.57	63.25	84.82	0.04568	0.16741	60
70	84.82	70.12	0.0121	0.493	82.60	2.028	23.90	61.92	85.82	0.05009	0.16701	70
80	98.76	84.06	0.0123	0.425	81.39	2.353	26.28	60.52	86.80	0.05446	0.16662	80
90	114.3	99.6	0.0125	0.368	80.11	2.721	28.70	59.04	87.74	0.05882	0.16624	90
100	131.6	116.9	0.0127	0.319	78.80	3.135	31.16	57.46	88.62	0.06316	0.16584	100
110	150.7	136.0	0.0129	0.277	77.46	3.610	33.65	55.78	89.43	0.06749	0.16542	110
120	171.8	157.1	0.0132	0.240	76.02	4.167	36.16	53.99	90.15	0.07180	0.16495	120
130	194.9	180.2	0.0134	0.208	74.46	4.808	38.69	52.07	90.76	0.07607	0.16438	130
140	220.2	205.5	0.0138	0.180	72.73	5.571	41.24	50.00	91.24	0.08024	0.16363	140

[1] Inches of Mercury below 1 atm.

stood before the work of Linde. When both liquid and gas are present in the same closed vessel, with free circulation, the temperature of both the gas and liquid are at the boiling point of the liquid at the pressure under which it exists.

When the pressure is increased over most liquids, there is an increase in the boiling temperature. Above a certain temperature the liquid

TABLE 2.8

CRITICAL TEMPERATURES OF CRYOGEN REFRIGERANTS

Cryogen	Critical Temp. °F
Acetylene	−103.5
Argon	−188.43
Carbon dioxide	+ 88.41
Helium	−450.31
Hydrogen	−399.91
Krypton	− 82.79
Neon	−379.75
Nitrogen	−232.87
Oxygen	−181.91
Xenon	+ 61.9

state ceases entirely and there is consequently no latent heat of vaporization. It remains a gas above that temperature, no matter how great the pressure exerted might be. This is the critical temperature.

The critical pressure is the pressure existent under conditions of critical temperature. It is possible to have a liquid condition at pressure above

TABLE 2.9

LOW TEMPERATURE GASES

	Boiling °F	Vol at 60 °F Atm Cu Ft per Lb	Freezing °F	Critical Temperature °F	Critical Pressure Psia
Oxygen	−297.3	11.86	−361.8	−181.9	730.6
Nitrogen	−320.5	13.55	−345.9	−232.9	492.2
Hydrogen	−423.0	188.15	−434.5	−399.9	190.8
Helium	−452.0	91.4	−453.5	−450.3	33.2

the critical pressure but when the critical temperature is attained no liquid condition of any specific gas can exist.

James Priestly had isolated gaseous and liquefied ammonia in 1798.

Oxygen was liquefied in 1877—nitrogen in 1883.

James Dewar liquefied hydrogen in 1898.

Heike Onnes liquefied helium in 1888.

Joule-Kelvin Cooling

By the combined investigations of Joule and Thompson (Lord Kelvin) a law of thermodynamics was forthcoming known as the Joule-Thompson or Joule-Kelvin law of gas expansion. In its simplest form the law states: When a gas expands without doing external work, its heat content remains unchanged.

The expansion value of most refrigeration systems operate under the conditions of the Joule-Thompson law of internal energy.

Joule-Kelvin or sometimes named Joule-Thompson cooling is very useful in the liquefaction of the cryogens. It affords a method of obtaining the necessary low temperature.

Carbon Dioxide

Carbon dioxide is probably the safest of all the primary fluids which have been used, with the exception of air and water. Its thermodynamic properties are not particularly desirable for refrigeration except cryogenic applications, since the low critical temperature, 87.8F, requires the condenser temperature to be held below this figure. Condensing pressure is exceedingly high and thus the heaviest construction is necessary for the compressors and condensers. This is not a completely unmitigated evil, since the high pressure results in a very low specific volume and, consequently, compressors may be made quite small for a given capacity. It is completely stable under all conditions of operation and is noncorrosive to all construction materials. Lubricating oils are almost completely insoluble in the liquid and this facilitates the separation and removal of lubricant from condensers and evaporators. This refrigerant is odorless and is not dangerous to life in any concentration below 4%, although in larger concentrations it acts as a suffocant. Leaks in carbon dioxide systems can be detected only by means of the bubbles produced in soap solutions. Carbon dioxide gas is compressed to approximately 1,200 psia in 2 or more stages. A typical operation of compression

TABLE 2.10

PROPERTIES OF CARBON DIOXIDE

Chemical symbol	CO_2
Triple point, °F	−69.9
Subliming point at atmospheric pressure °F	−109.3
Conditions at critical point	
Temperature °F	87.8
Pressure, psia	1070.0
Specific heat of vapor at constant pressure at 68 °F	0.1988
Specific heat of vapor at constant volume at 68 °F	0.1525
Latent heat of sublimation at 1 atm, Btu per lb	246.3
Suction pressure at 5 °F psia	332.0
Condensing pressure at 86 °F psia	1043.0
Toxicity, Underwriters Laboratories Grouping	5.0
Flammability	Noncombustible

TABLE 2.10A

PROPERTIES OF SATURATED CARBON DIOXIDE (REFRIGERANT 744)[1]

Temp °F t	Pressure Abs Psi p	Pressure Gage Atm/In. a-gp	Pressure Gage Psi gp	Volume Liquid Ft³/Lb v	Volume Vapor Ft³/Lb V	Density Liquid Lb/Ft³ 1/v	Density Vapor Lb/Ft³ 1/V	Enthalpy Above 32°F Liquid Btu/Lb h	Enthalpy Above 32°F Latent Btu/Lb L	Enthalpy Above 32°F Vapor Btu/Lb H	Entropy from 32°F Liquid Btu/Lb s	Entropy from 32°F Vapor Btu/Lb L/T	Entropy from 32°F Vapor Btu/Lb S	Temp °F t
-22	212.9	13.48	198.2	0.0155	0.4319	64.52	2.315	-24.78	126.7	102.0	-0.0533	0.2898	0.2365	-22
-20	220.6	14.00	205.9	0.0155	0.4166	64.34	2.401	-23.96	126.0	102.0	-0.0514	0.2867	0.2353	-20
-15	240.5	15.36	225.8	0.0157	0.3807	63.84	2.627	-21.88	124.1	102.2	-0.0467	0.2792	0.2325	-15
-10	261.7	16.80	247.0	0.0158	0.3482	63.25	2.872	-19.76	122.0	102.3	-0.0420	0.2716	0.2296	-10
-5	284.4	18.35	269.7	0.0160	0.3186	62.63	3.138	-17.61	119.9	102.3	-0.0372	0.2639	0.2267	-5
0	308.6	20.00	293.9	0.0161	0.2918	61.95	3.427	-15.41	117.7	102.2	-0.0324	0.2561	0.2237	0
5²	334.4	21.75	319.7	0.0163	0.2673	61.22	3.741	-13.16	115.3	102.1	-0.0275	0.2482	0.2207	5²
10	361.8	23.61	347.1	0.0165	0.2450	60.48	4.082	-10.87	112.8	101.9	-0.0226	0.2402	0.2176	10
15	391.0	25.60	376.3	0.0167	0.2245	59.73	4.454	-8.515	110.1	101.6	-0.0176	0.2321	0.2145	15
20	422.0	27.71	407.3	0.0170	0.2058	58.95	4.859	-6.102	107.3	101.2	-0.0126	0.2239	0.2113	20
25	454.8	29.94	440.1	0.0172	0.1886	58.14	5.303	-3.618	104.3	100.7	-0.0074	0.2154	0.2080	25
30	489.6	32.31	474.9	0.0175	0.1728	57.30	5.789	-1.049	101.2	100.1	-0.0021	0.2067	0.2046	30
35	526.4	34.81	511.7	0.0177	0.1581	56.41	6.323	1.604	97.77	99.38	0.0033	0.1978	0.2010	35
40	565.4	37.46	550.7	0.0180	0.1446	55.45	6.915	4.367	94.13	98.50	0.0088	0.1885	0.1973	40
45	606.5	40.26	591.8	0.0184	0.1321	54.41	7.571	7.251	90.21	97.46	0.0146	0.1788	0.1934	45
50	650.0	43.22	635.3	0.0188	0.1204	53.28	8.304	10.28	85.95	96.24	0.0205	0.1687	0.1893	50
55	695.9	46.34	681.2	0.0192	0.1095	52.05	9.132	13.40	81.29	94.78	0.0268	0.1580	0.1849	55
60	744.2	49.63	729.5	0.0197	0.0992	50.71	10.08	16.93	76.14	93.07	0.0335	0.1466	0.1801	60
65	795.1	53.09	780.4	0.0203	0.0894	49.14	11.18	20.66	70.35	91.01	0.0406	0.1342	0.1748	65
70	848.7	56.74	834.0	0.0211	0.0800	47.29	12.49	24.78	63.71	88.49	0.0484	0.1204	0.1688	70
75	905.1	60.57	890.4	0.0222	0.0708	45.05	14.13	29.50	55.83	85.33	0.0573	0.1045	0.1618	75
80	964.3	64.60	949.6	0.0235	0.0613	42.50	16.32	35.21	45.88	81.09	0.0679	0.0851	0.1530	80
85	1027.0	68.83	1012.3	0.0258	0.0500	38.76	20.00	43.18	31.29	74.47	0.0826	0.0575	0.1401	85

[1] Arranged by American Society of Refrigerating Engineers.
[2] Standard ton temperatures.

to 1,200 psia would be a first stage to 100 psia, a second stage to 300 psia and the final stage to 1,200 psia.

Nitrogen

Although there is approximately 21 million tons of nitrogen above every square mile of the earth, in the United States the commercial capturing of nitrogen from the air for agricultural and military uses did not succeed until World War I when the supply of Chilean nitrate was cut off to a critical point and the U.S. Department of the Army gambled on two possible processes that had advanced through the laboratory stage. The site chosen for the first plants was Muscle Shoals, Alabama.

TABLE 2.11

SOME PHYSICAL CHARACTERISTICS OF NITROGEN

Boiling point, normal, °F	−328.4
Freezing point, °F	−345.8
Critical temp, °F	232.87
Critical pressure at critical temp, psia	492.2
Vol. at 70°F and atmospheric, cu ft per lb	13.5
Molecular weight	28.0
Liquid density, normal, lb per cu ft	50.46
Latent heat, Btu per lb vaporization	85.8
Gas density at 70°F and 14.7 psia	0.072
Vapor pressure, psi	14.7
Specific heat C_p, Btu/lb R	0.49
Viscosity, Poises $\times 10^7$, 70°F	172.0
Liquid density at −335°F, lb per gal.	7.0
Triple point, °F	−346.03

The outcome of this effort was one process that was commercially successful, the other was prohibitive in cost. Today, from this 1919 Muscle Shoals pattern, scores of new and more advanced nitrogen and ammonia plants have been established throughout the world.

The cost of ammonia to the refrigeration industry has been reduced to less than one-third that prevailing prior to World War I. Nitrogen has become a common commodity on the American market not only for agricultural and military use but as a useful cryogenic liquid and gas in the quick freezing of foods and the low temperature treatment of both metallic and non-metallic essential manufacturing materials.

One process of food freezing consists of passing the blanched foodstuff to an insulated tunnel in which there is a pool of nitrogen that evaporates at −320°F. The blanched foodstuff at 180°F is moved above the insulated contained pool of liquid nitrogen which is evaporating at this −320°F temperature. When the foodstuff is exposed to this vapor, the temperature is lowered below 0°F in 4 to 6 min. This method is ex-

TABLE 2.12

THERMODYNAMIC CHARACTERISTICS OF NITROGEN
PRESSURE-ENTHALPY TABLE

| Temp °F | Pressure Psia | Volume | | Enthalpy | | |
		Liquid Cu Ft/Lb	Vapor Cu Ft/Lb	Heat of Liquid Btu/Lb	Latent Heat Btu/Lb	Total Heat Btu per Lb
−320.4	14.7	0.01985	3.47	0.0	85.64	85.64
−315.0	20.8	0.02018	2.51	2.57	84.02	86.59
−310.0	27.5	0.02053	1.93	5.08	82.32	87.54
−305.0	36.0	0.02090	1.50	7.61	80.51	88.13
−300.0	46.5	0.02129	1.19	10.18	78.59	88.77
−295.0	59.0	0.02171	0.94	12.78	76.54	89.32
−290.0	73.9	0.02215	0.76	15.41	74.36	89.77
−285.0	91.2	0.02263	0.62	18.07	72.02	90.09
−280.0	111.3	0.02316	0.51	20.8	69.45	90.25
−275.0	134.3	0.02326	0.42	23.63	65.88	90.20
−270.0	160.7	0.02444	0.35	26.56	63.37	89.93
−265.0	190.5	0.02520	0.29	29.59	59.85	89.44
−260.0	224.1	0.02604	0.24	32.75	56.01	88.76
−255.0	261.6	0.02700	0.20	36.14	51.64	87.82
−250.0	304.0	0.02820	0.17	39.80	46.82	86.62
−245.0	350.9	0.02997	0.135	43.80	40.96	84.77
−240.0	402.8	0.03238	0.110	48.09	33.75	81.84
−235.0	460.4	0.03623	0.083	53.87	23.78	77.65
−232.42	492.9	0.05092	0.059	66.19	00.0	66.19

TABLE 2.12A

THERMODYNAMIC CHARACTERISTICS OF NITROGEN
AT TEMPERATURES OF SATURATION

Pressure Psia	Temp °F	Enthalpy Btu/Lb	Entropy Btu/Lb/R	Volume Cu Ft/Lb
10	−326.2	84.6	0.6333	4.95
15	−320.1	85.7	0.6144	3.402
20	−315.8	86.52	0.6988	2.604
25	−312.3	87.13	0.5898	2.111
30	−309.0	87.62	0.5811	1.782
40	−303.6	88.39	0.5670	1.359
50	−298.9	88.93	0.5559	1.102
60	−294.7	89.34	0.5471	0.9259
80	−287.9	89.91	0.5328	0.7008
100	−282.7	90.20	0.5209	0.5627
150	−272.0	90.66	0.4675	0.3728
200	−263.7	89.33	0.4778	0.2736
250	−256.5	88.15	0.4605	0.2111
300	−250.4	85.88	0.4441	0.1677
400	−240.3 critical	81.76	0.4078	0.1104
500	−230*	80.39	0.3968	0.0821
600	−230*	55.88	0.2866	0.0356
800	−230*	51.51	0.2621	0.03134
1000	−230*	49.15	0.2466	0.02930
1500	−230*	45.92	0.2235	0.02639

* Approximately.

pensive when the nitrogen cooling is carried from blanching temperature on down to chilling and finally to —5° to —10°F. This complete nitrogen cooling may cost, from blanching to freezing, 7 to 10 cents per pound of product for the entire operation.

Other proposed experimental ultra-rapid freezing operations follow a sequence of cooling from the blanching temperature of 180° to 205°F to subsequent air blast or ice water immersion to 32°F, thence by some secondary refrigerant immersion to about 15°F and complete the cycle by nitrogen gas to —10°F. This procedure reduces the overall cost, although the labor involved is somewhat greater than for reducing the temperature from that of blanching to the final quick frozen state by nitrogen alone.

The three-step procedure does have an additional advantage for the quick freezing of berries and fruits in that the secondary refrigerant can be a low viscous invert sugar which at 15°F flows freely. The resultant coating of sugar glaze will resist the possible bleaching action of a dry gas at a low temperature.

Nitrous Oxide

Liquid nitrous oxide quick freezing was first introduced in Frankfurt, Germany in 1946 by I.C. Farbenindustrie. The equipment used by the Germans during World War II was later discovered by the Allied Forces when they took possession of the industries of Germany in the postwar reparations program.

The present barrier to the expansion of the commercial quick freezing by nitrous oxide is its high cost. Only when used as a closed refrigerant system can nitrous oxide be used economically.

TABLE 2.13

NITROUS OXIDE PHYSICAL PROPERTIES

Molecular weight	44.0
Boiling point, °F	—127.237
Latent heat of vaporization, Btu/lb	161.78
Sensible heat, Btu/lb	40.00
Total heat to 70°F	201.78
Liquid density at boiling point, lb/cu ft	76.54
Gas density at 32°F lb/cu ft	0.1146
Specific volume, cu ft/lb	8.711
Specific heat ratio at 70°F	1.26
Specific heat at constant pressure	0.2095
Specific heat at constant volume, 70°F	0.1609
Color	Clear
Odor	None
Toxicity	Nontoxic

SECONDARY REFRIGERANTS

A secondary refrigerant such as brine or other liquids, is used to convey heat from the point of the load to the evaporator. Included in this classification are glycols, alcohols, liquids with a low temperature freezing point. This is in contradistinction to the gaseous or liquid refrigerant which transfers heat or cold directly.

Disadvantages of Direct Expansion System

The disadvantages of the direct expansion system should not be overlooked. First, any leak that might occur in the direct expansion coils will discharge directly into the refrigerator, and if the material in storage is not in sealed tins or cases, the refrigerant odor will be immediately absorbed by the products in storage.

Second, the direct expansion system requires a large amount of high-pressure refrigerant piping and fittings. Consequently, the loss of refrigerant in such a system may be excessively high. This disadvantage is not seriously experienced in plants where the refrigerators must be reached by long lines installed either underground or in obscure passageways.

Third, the direct expansion system has little hold-over capacity. When the compressor stops there is no reserve of cooling medium that can be pumped to coils to maintain the desired temperatures.

Brine Distribution System

The brine distribution system is used most extensively for ice-making and for long-distance transfer of "cold." In general, it consists of submerging the primary refrigerant piping beneath the brine, cooling this brine to the desired point, then circulating the chilled brine through pipes in the room where refrigeration is desired.

Brine Coolers

In ice-making plants, the brine cooler is usually built-up of coils of pipe arranged in vertical rows between each row of ice cans. For cold-room work, either a double-pipe coil or a shell-and-tube brine cooler is used however.

The shell-and-tube coolers used for cold rooms are generally of the multipass design, the brine passing back and forth through the several passes while the refrigerant occupies the space between the coils.

TABLE 2.14

THERMODYNAMIC PROPERTIES OF SATURATED NITROUS OXIDE

Temperature °F	Absolute Pressure Psi	Volume, Cu Ft/Lb		Enthalpy, Btu/Lb			Entropy, Btu/(Lb) (°R)		
		Satd. Liquid	Satd. Vapor	Satd. Liquid	Vapori-zation	Satd. Vapor	Satd. Liquid	Vapori-zation	Satd. Vapor
-127.237	14.696	0.01310	5.069	0.00	161.72	161.72	0.0000	0.4864	0.4864
-125	15.82	0.01313	4.758	0.93	161.14	162.07	0.0024	0.4815	0.4839
-120	18.60	0.01322	4.141	3.02	159.83	162.85	0.0079	0.4705	0.4784
-115	21.74	0.01331	3.586	5.14	158.48	163.62	0.0135	0.4597	0.4732
-110	25.28	0.01340	3.107	7.30	157.09	164.39	0.0191	0.4492	0.4683
-105	29.25	0.01349	2.704	9.48	155.67	165.15	0.0247	0.4389	0.4636
-100	33.68	0.01358	2.374	11.69	154.21	165.90	0.0304	0.4287	0.4591
-95	38.62	0.01368	2.096	13.92	152.72	166.64	0.0360	0.4188	0.4548
-90	44.09	0.01378	1.856	16.17	151.20	167.37	0.0418	0.4090	0.4508
-85	50.14	0.01388	1.646	18.46	149.64	168.10	0.0475	0.3994	0.4469
-80	56.79	0.01398	1.463	20.75	148.05	168.80	0.0534	0.3899	0.4433
-75	64.10	0.01409	1.302	23.07	146.42	169.49	0.0593	0.3806	0.4399
-70	72.09	0.01420	1.159	25.42	144.74	170.16	0.0654	0.3714	0.4368
-65	80.80	0.01432	1.0396	27.81	143.01	170.82	0.0717	0.3623	0.4340
-60	90.29	0.01444	0.9388	30.22	141.23	171.45	0.0782	0.3533	0.4315
-55	100.58	0.01456	0.8515	32.68	139.38	172.06	0.0847	0.3444	0.4291
-50	111.71	0.01469	0.7751	35.18	137.45	172.63	0.0912	0.3355	0.4267
-45	123.73	0.01482	0.7084	37.73	135.45	173.18	0.0978	0.3266	0.4244
-40	136.68	0.01495	0.6483	40.30	133.39	173.69	0.1044	0.3178	0.4222
-35	150.60	0.01509	0.5929	42.90	131.27	174.17	0.1109	0.3091	0.4200
-30	165.54	0.01524	0.5416	45.49	129.11	174.60	0.1173	0.3005	0.4178
-25	181.53	0.01539	0.4938	48.06	126.93	174.99	0.1236	0.2920	0.4156
-20	198.62	0.01555	0.4503	50.60	124.74	175.34	0.1296	0.2887	0.4133
-15	216.85	0.01572	0.4111	53.10	122.53	175.63	0.1355	0.2755	0.4110
-10	236.28	0.01589	0.3759	55.58	120.29	175.87	0.1411	0.2675	0.4086
-5	256.98	0.01607	0.3444	58.03	118.03 ·	176.06	0.1465	0.2596	0.4061

0	278.97	0.01625	0.3162	60.47	115.73	176.20	0.1518	0.2518	0.4036
5	302.31	0.01645	0.2908	62.91	113.39	176.28	0.1570	0.2440	0.4010
10	327.03	0.01666	0.2675	65.30	111.00	176.30	0.1620	0.2363	0.3983
15	353.21	0.01688	0.2463	67.72	108.54	176.26	0.1669	0.2287	0.3956
20	380.88	0.01711	0.2270	70.15	106.01	176.16	0.1718	0.2210	0.3928
25	410.07	0.01735	0.2093	72.61	103.38	175.99	0.1767	0.2133	0.3900
30	440.87	0.01761	0.1931	75.11	100.63	175.74	0.1817	0.2055	0.3872
35	473.33	0.01789	0.1781	77.68	97.74	175.42	0.1868	0.1976	0.3844
40	507.51	0.01819	0.1641	80.32	94.68	175.00	0.1920	0.1895	0.3815
45	543.46	0.01851	0.1512	83.06	91.43	174.49	0.1974	0.1812	0.3786
50	581.25	0.01886	0.1391	85.89	87.95	173.84	0.2029	0.1726	0.3755
55	620.97	0.01925	0.1278	88.85	84.21	173.06	0.2086	0.1636	0.3722
60	662.69	0.01968	0.1171	91.94	80.16	172.10	0.2145	0.1542	0.3687
65	706.50	0.02017	0.10701	95.16	75.75	170.91	0.2203	0.1444	0.3647
70	752.49	0.02073	0.09741	98.52	70.89	169.41	0.2263	0.1338	0.3601
75	800.78	0.02138	0.08818	102.02	65.47	167.49	0.2322	0.1224	0.3546
80	851.51	0.02218	0.07921	105.71	59.28	164.99	0.2382	0.1098	0.3480
85	904.83	0.02321	0.07029	109.78	51.98	161.76	0.2446	0.0954	0.3400
90	960.98	0.02465	0.06113	114.72	42.81	157.53	0.2523	0.0779	0.3302
95	1020.24	0.02731	0.05009	122.02	28.55	150.57	0.2641	0.0515	0.3156
97.581	1052.17	0.03540	0.03540	136.35	0	136.35	0.2890	0	0.2890

TABLE 2.15

COMPARATIVE REFRIGERANT AND CRYOGEN PERFORMANCE PER STANDARD TON (86°F CONDENSATION, 5°F SUCTION) OF REFRIGERANTS MOSTLY USED IN COLD AND FREEZER WAREHOUSING

Refrigerant	No.	Evaporator Pressure Psig	Condensing Pressure Psig	Refrigerant Circulated Lb/Min	Net Refrigerating Effect Btu/Lb	Coefficient of Performance	Horsepower per Ton	Compressor Discharge Temp °F	Compression Ratio
Nitrous oxide	744A	294.3	922.3	2.35	85.2	3.60	1.310		3.03
Carbon dioxide	744	317.5	1031.0	3.62	55.5	2.56	1.840	151	3.15
Ammonia	717	19.6	154.5	0.422	474.4	4.76	0.989	210	4.94
Methyl chloride	40	6.5	80.0	1.33	150.2	4.90	0.962	172	4.48
Dichloroethylene	1130	28.3	15.8	1.75	114.3	4.83	0.973		8.42
Trichloroethylene	1120	29.6	26.2	2.18	91.7	4.82	0.980		11.65

Heat Transfer in Brine Coolers.—Brine coolers necessarily involve two steps in the cooling process whereas the direct system entails only one. The brine must be first cooled by the refrigerant; then the brine will, in turn, cool the refrigerator. These two exchanges must necessarily require a greater difference of temperature between the refrigerant and the ultimate refrigerator than would the one exchange of the direct expansion system. This greater difference of temperature requires that lower suction pressures shall be maintained for the same degree of cooling when the indirect system is used, and lower suction pressures necessarily require a greater horsepower input per ton of refrigeration produced.

Fortunately, for the development of indirect systems, the great improvement in brine-cooling equipment has reduced this margin to a minimum. Brine coolers are now available that are sold on a guarantee of only 2°F difference in temperature between the brine and boiling refrigerant. Whereas the extra cost of rapid circulation of the brine in such systems must be charged to the system, it, nevertheless, has reduced the wide gap in horsepower requirements between the direct and indirect expansion methods of refrigerating rooms.

For central station refrigeration, the brine circulation method has been found to be very satisfactory. The distribution of refrigerant through long pipe lines has never been looked upon with great favor due to hazard involved in leaks from the high-pressure refrigerant lines. But relatively low pressures can be employed when brine is the cold-carrying agent, and a leak might be of considerable magnitude without causing undue damage since the brine is relatively cheap and practically odorless.

The tanks used for the brine can be made of steel, concrete or wood; in any case, they should be amply insulated, then sealed with a positive vapor barrier to keep the insulation dry. It usually requires 6 to 12 in. of built-up insulant with a k factor of not less than 0.30, this to reduce the loss of cold through the walls.

Characteristics of Brines.—When a small quantity of one substance is dissolved in another, the resulting solution has a freezing point which is usually lower than that of the second substance. As most of the first substance is dissolved, the freezing point is further depressed until a minimum freezing temperature is attained. This is called the eutectic point of temperature and composition of the solution is known as the eutectic. Any further addition of the substance to the solution will not lower its freezing point; instead when the temperature is lowered to the eutectic point the extra added substance will solidify or pass out of solution and the remaining liquid will have the eutectic composition. If the temperature is lowered beyond the eutectic point, the entire mixture will freeze solid.

Sodium Chloride Brines.—Water solutions of sodium chloride, frequently known as "sodium brines," are very widely employed, since common salt is very low in price and is readily procured. The eutectic temperature in water solution is $-6.0°F$, which covers the range of ice-making and of the majority of cold storage installations. Specific heat and thermal conductivity are high and the viscosity is low enough to minimize the power required for circulation. In packing houses and food plants where brine may accidentally leak or spill upon foodstuffs, it is desirable to use salt brine instead of the unpleasant tasting calcium or magnesium chloride brines. Commercial common salt is frequently acid in character because of the presence of impurities. Careful neutralization of the brine is thus necessary to prevent corrosion of the brine system and piping.

Calcium Chloride Brines.—The eutectic temperature of the water solution of calcium chloride is $-59.8°F$ and the eutectic composition contains 29.8% of the anhydrous material. Thus, calcium brines may be used successfully at any temperature down to approximately $-55°F$. The specific heat is somewhat less than equivalent common salt brine and the viscosity is greater. For hold-over tanks in which the brine is partially frozen to store refrigerating effect, calcium brine is particularly desirable, as the solid salt does not precipitate out until $-59.8°F$ is attained. Freshly made calcium chloride brine is somewhat alkaline and should be neutralized if it is to be used in contact with galvanized iron. Muriatic acid may be used as a neutralizer, but it requires careful chemical control since an excess will cause vigorous attack upon all metal parts. The simplest method of neutralizing excessive alkalinity is to pass a stream of carbon dioxide gas into the agitated solution until no pink color is produced by adding a drop of phenolphthalein test solution to a teaspoon of the brine.

Corrosion in Brine Systems.—Brine systems are usually constructed of iron or steel because of the relatively low cost and convenience of these construction materials. In the presence of water, air and soluble salts, severe rusting and corrosion may occur if steps are not taken to minimize chemical action. At times, some wild claims have been made that certain brines are noncorrosive to metals; however, it may be stated that the chlorides of either sodium, calcium or magnesium will cause corrosion of steel and iron parts in the presence of water and air. The presences of acidic impurities in brine salts greatly accelerates corrosion.

When only iron or steel parts are present in the system, it is desirable to maintain a high degree of alkalinity in the brine. However, zinc alloys and galvanized parts are rapidly attacked by alkali; thus, the brine in an ice tank must be maintained in an approximately neutral condition to prevent attack on the galvanized ice cans. Some operators prefer submerging a bar of magnesium.

The pH Scale.—To measure the exact degree of acidity and alkalinity, a special scale has been devised, which is known as the hydrogen ion concentration or the pH scale. On this scale, a value of 7.0 indicates that the solution is balanced between acidity and alkalinity; thus, it is neutral. A pH value less than 7.0 indicates that the acidic character predominates, whereas above 7.0 the alkaline property is greatest. The scale is logarithmic in character and a change of one unit indicates that a ten-

TABLE 2.16

ACIDITY AND ALKALINITY OF SOLUTION IN RELATION TO PH

Value of pH	Relative Acidity of Solution	Relative Alkalinity of Solution
4	1000	0.001
5	100	0.01
6	10	0.1
7	1	1
8	0.1	10
9	0.01	100
10	0.001	1000

fold change in both acidity and alkalinity has occurred. For instance, at pH 7.0 the acidity and alkalinity are exactly equal, but at pH 8.0 the acidity has decreased tenfold whereas the alkalinity has increased tenfold. The following table of values will show these relations for varying pH values.

Miscellaneous Suggestions for Good Plant Practice

(1) In all systems steps should be taken to prevent, in so far as possible, the agitation of brine in contact with air; all inlets should be sealed. In general, the elimination of dissolved oxygen will largely prevent corrosion difficulties and any efforts in this direction will be well repaid.

(2) It is suggested that weighted test pieces of galvanized and bar iron or steel be hung in the brine system as a convenient method of conservation.

(3) If brine is not treated, it should be maintained slightly alkaline (pH about 8.5). The use of phenolphthalein as an indicator of alkalinity is recommended. A faint pink coloration should be maintained, if

necessary, by the addition of a small amount of caustic soda or milk of lime.

(4) Contact of dissimilar metals and brines should be avoided as far as practicable. The dichromate treatment will in most cases minimize this type of corrosion.

Calcium chloride brines may be successfully employed at temperatures as low as —55°F.

ATMOSPHERIC AIR

Atmospheric air as normally found in nature is a mixture of dry air and water vapor. The percentage composition of dry air is approximately 78.03 nitrogen, 20.99 oxygen, 0.94 argon, 0.03 hydrogen and some rare gases such as xenon and krypton 0.01. For practical purposes, air by volume is considered made up of 79% nitrogen and 21% oxygen or by weight 76.8% nitrogen and 23.2% oxygen.

The water vapor in air does not show as a fog. When fog appears, it is as a supersaturated cloud. Clear air or invisible air may have as high as 100% saturation but when it exceeds 100% saturation, the moisture becomes visible as a fog.

The computations and relationships of the vapor air mixtures constitute the study of psychrometrics.

Atmospheric air exists at a pressure called barometric. This barometric pressure is made up of partial pressure exerted by all the gases of the mixture. The sum total barometric pressure for practical purposes is considered as made up of three partial pressures: oxygen, nitrogen and water vapor.

TABLE 2.17

PROPERTIES OF DRY AIR

Temperature °F	Volume in Cu Ft per Lb	Heat Content Btu per Lb Dry Air
50	12.84	12.005
55	12.97	13.206
60	13.10	14.406
65	13.22	15.607
70	13.35	16.807
75	13.47	18.008
80	13.60	19.208
85	13.73	20.41
90	13.85	21.61
95	13.90	22.81
100	14.10	24.01
105	14.23	25.21
110	14.36	26.41
115	14.48	27.61
120	14.61	28.81

TABLE 2.18

PROPERTIES OF SATURATED AIR

Temp °F	Volume in Cu Ft per Lb	Btu per Lb	Grains Moisture per Cu Ft	Grains Moisture per Lb Dry Air	Latent Heat Btu per Lb Dry Air
30	12.40	10.92	1.8	24.0	3.71
35	12.55	13.00	2.2	29.1	4.00
40	12.70	15.21	2.8	35.5	5.62
45	12.85	17.63	3.3	46.0	6.84
50	13.00	20.26	4.10	53.42	8.26
55	13.16	23.18	4.96	64.35	9.97
60	13.33	26.40	5.80	77.28	11.99
65	13.50	29.99	6.85	92.49	14.38
70	13.69	33.99	8.06	110.30	17.19
75	13.88	38.49	9.45	131.16	20.48
80	14.09	43.55	11.04	155.61	24.35
85	14.31	49.28	12.87	184.17	28.87
90	14.54	55.75	14.94	217.27	34.14
95	14.80	63.08	17.28	255.81	40.27
100	15.08	71.44	19.94	300.70	47.43
105	15.38	80.99	22.94	352.9	55.78
110	15.72	91.92	26.31	413.6	65.51
115	16.09	104.46	30.01	484.3	76.85
120	16.51	119.02	34.34	567.1	90.21

TABLE 2.19

REQUIRED CUBIC FEET OF AIR AT HIGHER ELEVATIONS AS COMPARED TO SEA LEVEL

Elevations	Requirement For Same Duty as at Sea Level, Cu Ft
0	100
1000	103
2000	107
4000	114
6000	123
8000	131
10000	141

MEASUREMENT OF HUMIDITY AND TEMPERATURE IN COLD AND FREEZER STORAGE ROOMS

Measuring Temperature Only

Dry Bulb Temperatures.—Under normal conditions, all room temperatures are taken with a dry bulb thermometer, that is, a thermometer with a glass mercury bulb exposed to the atmosphere. For most engineering work in North America and Great Britain the unit selected is the Fahrenheit degree, and the starting point is 32°F below the freezing point of water (0°F). The drying effect, the cooling effect, or the comfort of a room are not entirely dependent upon the temperature of the room. The movement of the air, the moisture content, and the radiant heat impact from the walls and ceilings will also influence all three of these factors. The same temperature maintained in a room in which the moisture content is varied from a high value to a low value may pass from a point of human comfort to one of decided chilliness. During this entire change, the thermometer might remain at a standstill, the amount of moisture or the air movement only changing.

Measuring Moisture Content Only

The factor of air dilution with moisture is known as humidity. Air dilution can be measured in grains of water vapor per cubic foot of air.

TABLE 2.20

HEAT TO REMOVE IN BTU PER CU FT IN COOLING AIR FOR COLD AND FREEZER STORAGE

For Storage Room at Temp °F	Temperature of Incoming Air			
	50°F		100°F	
	Relative Humidity			
	70% RH	80% RH	50% RH	60% RH
30	0.58	0.66	2.95	3.35
25	0.75	0.83	3.14	3.54
20	0.91	0.99	3.33	3.73
15	1.06	1.14	3.51	3.92
10	1.19	1.27	3.64	4.04
5	1.34	1.42	3.84	4.27
0	1.48	1.56	4.01	4.43
−10	1.73	1.81	4.31	4.74
−20	2.01	2.09	4.66	5.10
−30	2.29	2.38	4.90	5.44

When air moisture content is measured on this basis, it is called *Absolute Humidity*. In measuring absolute humidity no mention is made of the temperature. A cubic foot of air at 32°F containing 10 grains of water will have the same absolute humidity as a cubic foot of air at 120°F containing 10 grains of water. In each case its absolute humidity would be said to be 10 grains per cu ft.

Measuring Moisture and Temperature Together

It is necessary to measure moisture and temperature together to determine humidity. If two thermometers are mounted side by side, and a wet gauze placed over one of them, then the pair moved about in the room atmosphere, unless the air is saturated with moisture, the temperature reading of the wet bulb thermometer will be depressed and the depression that is noted in relation to the dry bulb thermometer reading will indicate the humidity. This arrangement makes up what is called a simple sling psychrometer—one of the most accurate methods of humidity measurement.

Room Volume Cu Ft	Air Changes per 24 Hr	Room Volume Cu Ft	Air Changes per 24 Hr
200	44.0	500	26.0
1000	17.5	2000	12.0
5000	12.0	4000	8.2
6000	6.5	8000	5.5
10,000	4.9	20,000	3.5
30,000	2.7	40,000	2.3
50,000	2.0	100,000	1.4

Temperature Effect and Relative Humidity

Cold air will not absorb as much moisture as warm air. At 32°F a cubic foot of air will hold 2 grains of water vapor. At 50°F it will hold 4 grains of vapor. At 100°F it will hold 19.5 grains of water vapor. These values are known as the saturation points at these temperatures. If water vapor to the amount of 1 grain is present in a cubic foot of air at 32°F, that cubic foot of air would be 50% saturated. If now that cubic foot of air were 50°F, it would be only 1 in 4 or 25% saturated, and if it were 100°F it would be 1 in 19.5 or just slightly over 5% saturated. This degree of saturation is referred to as its *Relative Humidity*.

When the air contains all of the water vapor it will hold it is saturated or it has a relative humidity of 100%. When air is saturated it has practically no drying capacity. For drying then we must have air of low humidity. Air at 32°F containing 2 grains of water per cubic foot is saturated. It would have no drying capacity. Heat this same air to 50°F and its relative humidity would drop to 50% and it could absorb 2 more grains of water. Continuous heating of this cubic foot of air to 100°F and it would have 5% humidity and it would absorb 17.5 gr more water than at 32°F.

Dew Point

Reverse this experiment, that is, start with a cubic foot of air at 100°F containing 19.5 grains of water vapor or with a relative humidity of 100%. If the air is cooled below 100°F it will not be able to retain all of its moisture and the excess water will condense. The condensing point is known as the *Dew Point*.

HUMIDITY CONTROL

There are several expressions the operator meets in working in cold and freezer rooms. Humidity control of the freezer rooms is of less concern than for chill and cooler storage space but must not be ignored even at 0°F. Humidity refers to the amount of water vapor within the air of a definite space. It is a most critical factor in storage warehousing. For example, in the cold storage of meats, too high humidity creates excessive mold growth, too low humidity causes serious dehydration. Either practice causes deterioration of meat.

Absolute humidity is the weight of water vapor per unit volume, grains per cubic foot or grams per cubic centimeter of air.

Dry bulb temperature of a gas or mixture of gases is indicated on an accurate thermometer or pyrometer when there is no heat flow to and from the bulb or the couple.

Wet bulb is the temperature indicated by a thermometer which has a wetted bulb over which air is flowing.

Dewpoint is the temperature at which condensation of water vapor begins for an existing state of humidity and pressure when moist air is cooled at constant pressure.

Relative humidity is the ratio of the actual partial pressure of the water vapor in a space to the saturation pressure of pure water at the same temperature.

Specific humidity is the ratio in a mixture of vapor and air of the weight of water vapor associated with 1 lb of dry air.

Saturated air connotes a mixture of dry air and saturated water vapor in which the partial pressure of water vapor is equal to the vapor pressure at the existing temperature.

Designing for Odor and Taste Control in
Cold and Freezer Storage Warehouses

Builders of freezer storage installations have a great responsibility in odor and taste control in building design. Odors may move about through porous walls, cracks, doors, windows, elevator shafts, transportation and ventilating systems, on living organisms including man and by flowing liquids and gases. Refrigeration and air conditioning engineers designing and maintaining cold storage rooms, clean rooms, hospitals and miscellaneous food storage facilities recognize that it is often more sensible to contain such odors at the source than to permit their widespread distribution and dissemination.

Odors penetrating into cooler and freezer storage rooms soon contaminate the more sensitive foods (not properly packaged), especially those containing oils and fats, and usually lower the grade level of the product stored by producing an off-taste in the food. The lower the freezer storage room temperature, the less will be the odor migration since it lowers the vapor pressure of the penetrating gas carrier.

The olfactory system of human beings varies widely between individuals. Even in persons with a high degree of sensitivity to odors, continued presence within an odorous region may bring about a form of odor fatigue that dulls the sensitivity to that particular smell. In inspecting empty storage rooms for telltale odors and suitability for foodstuff storage, such inspection is best made by individuals entering from currently odor-free fresh air locations. Freezer storage rooms have limited problems of odor control since the frozen goods have a very low vapor pressure. Preparation for freezer storage does require the extraction of the latent heat of fusion of the water content of the product to be stored plus a small amount of heat of the solid. While odor migration is greatly reduced by the lower temperatures within freezer rooms, it must be recognized that in bringing the produce down to the predetermined room temperature level, migratory odors may penetrate the food stuffs.

There are also many perishable food items that have a freezing range extending much below the freezer room temperature. There will be odor movements into this unfrozen fraction of the food in question.

PSYCHROMETRIC CHARTS

Two psychrometric charts are presented herewith as designed by W. R. Woolrich. Figure 2.1 is for air-conditioning temperature from 40° to 120°F. Figure 2.2 is for refrigeration applications from —40° to +40°F. They show the interrelation between heat content Btu per lb of dry air and the pressure of water vapor psi, both of these plotted against dry bulb temperature in degrees above Fahrenheit. It also shows wet bulb temperature and the volume in cubic feet of dry air per pound of dry air at the different dry bulb temperatures.

ACTIVATED CHARCOAL ADSORPTION OF WAREHOUSE ODORS

The phenomenon of the physical condensation of a gas on charcoal surfaces is classified as adsorption in contrast to absorption. Highly selective charcoal for gas masks and similar air filtering applications is usually prepared from coconut shells and peach kernels. With the increase in demand for activated charcoal, many new sources including coal and wood charcoal have been developed.

To activate the charcoal, it is exposed to heating in a neutral atmosphere then the carbon particles exposed to a high temperature oxidizing

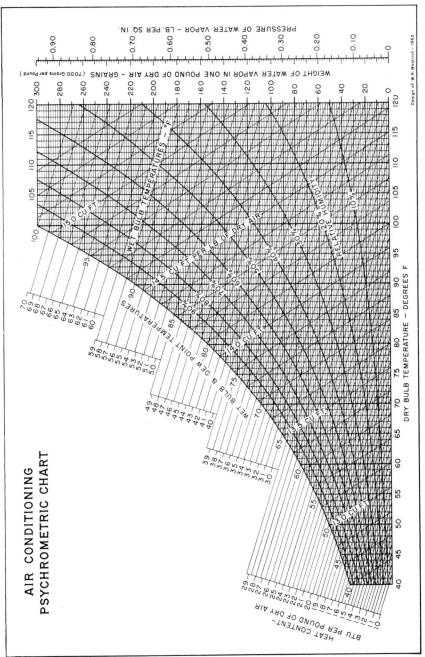

FIG. 2.1. AIR CONDITIONING PSYCHROMETRIC CHART

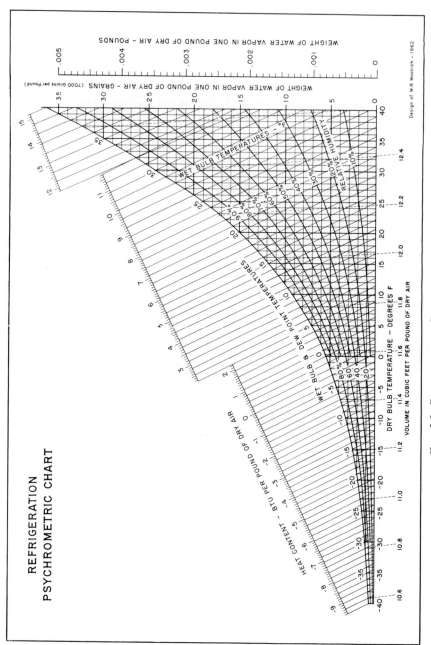

REFRIGERATION
PSYCHROMETRIC CHART

Fig. 2.2. Refrigeration Psychrometric Chart

process to remove unwanted substances within the base carbon material. This leaves the carbon cellular surface greatly increased within its exposed interstices. Activated charcoal is commercially sold in both cylindrical and flat plate cans. Its capacity to adsorb gaseous odors is so great that reactivation is not usually economical to the individual user. Most suppliers will offer instructions of high temperature treatment to revive the charcoal surfaces to their original adsorption capacity, but the individual usually finds it more economical to purchase a new charge in a replacement cannister.

ODORS AND VENTILATION AND FRESH AIR

Fresh air ventilation is used to dilute the inside air to bring the levels of obnoxious odors to an acceptable level. Care must be taken to assure the available so-called fresh air is not already polluted and thus unacceptable as a diluent.

Most ambient air in highly congested areas will be found to contain finely divided carbon, lint, spores and dust particles that are solids, then there may be a host of gaseous products in the same supply of air from chimneys, automobiles and industrial processes. It is often advisable to use a preliminary filter of the lesser efficient type; then feed into the activated carbon, or an electrostatic filter, the pre-filtered air of a better quality than the ambient atmosphere.

BIBLIOGRAPHY

AIR MOVING AND CONDITIONING ASSOC. 1962. Standard test code for air moving devices. AMCA Bull. *210.*

ALBRIGHT, L. F., and MARTIN, J. J. 1952. Thermodynamic properties of chlorotrifluoromethane (Freon 13). Ind. Eng. Chem. *44,* 188–189.

ASHLEY, C. M. 1942. The heat transfer of evaporating Freon. Refrig. Eng. *42,* No. 2, 89–95.

ASHLEY, C. M. 1950. Carrene-7, a new refrigerant. Refrig. Eng. *58,* 553–557, 605–606.

BAKER, M. 1956. Heat transfer rates from horizontal tubes to dichlorodifluoromethane. Refrig. Eng. *64,* 35–37.

BRYAN, W. L., and QUAINT, G. W. 1951. Heat transfer coefficients for horizontal tube evaporators. Refrig. Eng. *59,* No. 1, 67–72.

CONSLEY, J. C. 1938. Heat transfer in ammonia shell-and-tube brine coolers as affected by operating conditions. Refrig. Eng. *45,* No. 6, 409–415.

EDWARDS, H. D. 1944. Hydrogen refrigerants in low temperature fields. Refrig. Eng. *48,* 117–125, 144, 146, 148, 150, 152.

FISKE, D. L. 1949. Low temperature Freon refrigerants. Refrig. Eng. *57,* 336–339.

HUGHEY, T. M. 1946. Freon and ammonia as refrigerants. J. Milk Technol. 9, 220–224.

JOHNSON, R. C., JR., and CHADDOCK, J. B. 1964. Heat transfer and pressure drop of refrigerants evaporating in horizontal tubes. ASHRAE Trans. 70, Paper 1875, 163–172.

JONES, R. L., and CHAMBERLAIN, J. R. 1942. Recent developments in large ammonia absorption systems. Refrig. Eng. 43, 25–31.

JONES, W. 1941. Cooler and condenser heat transfer with low pressure Freon refrigerant. Refrig. Eng. 41, No. 6, 413–418.

KERKA, W. F., and HUMPHREYS, C. M. 1956. Temperature and humidity effect on odor perception. ASHRAE Res. Rept. 1587. ASHRAE Trans. 62, 531–533.

LOPICCOLO, T. 1951. Low temperature refrigeration systems. Refrig. Eng. 59, 239–242, 296–297.

MAY, J. W. 1956. Selecting the right air filter carefully. Refrig. Eng. 64, No. 3, 57.

MCCALL, N. H. 1954. Comparison of refrigerants. World Refrig. 5, 316–319, 344, 376–379, 497–500, 506.

MCKENNA, F. W. 1945. Design and operating characteristics of ammonia, Freon-12, Freon-22, and methyl chloride. Ice and Refrig. 108, No. 6, 35–38; 109, No. 1, 41–42, 44, 46, 48, 66.

MUNKELT, F. H. 1948. Air purification and deodorization by use of activated carbon. Refrig. Eng. Application Data 42. Am. Soc. Refrig. Eng. 1948.

NATL. BUR. STD. 1923. Tables of thermodynamic properties of ammonia. US Dept. Comm. Circ. 142, Washington, D.C.

NATL. BUR. STD. 1944. Solid carbon dioxide (Dry Ice). US Dept. Comm. Letter Circ. LC-763, Rev. Washington, D.C.

NATL. FIRE PROTECTION ASSOC., Boston, 1960. Fire hazard properties of flammable liquids, gases and volatile solids. NFPA Standard 325.

OWENS, R. G. 1945. Preparation of tables and charts for refrigerants. Refrig. Eng. 50, 225–228.

QUERMANN, C. C. 1945. Large ammonia compressors. Refrig. Eng. 50, 315–317, 349.

REICH, G. T. 1945. Carbon dioxide in the food industry. Food Inds. 17, No. 8, 84–86, 204; No. 9, 102–104.

ROWLEY, F. B., and JORDAN, R. C. 1938. Air filter performance as affected by kind of dust, rate of dust feed, and air velocity through filter. ASHRAE Res. Rept. 1094. ASHVE Trans. 44, 415–437.

SAWENS, R. H. 1947. Calcium chloride and sodium chloride refrigeration brines. Refrig. Eng. 54, No. 5, Refrig. Eng. Appl. Data Section 40, 1–5.

THOMPSON, R. J. 1937. The technical aspects of Freon refrigerants. Refrig. Eng. 33, 225–230.

WOOLRICH, W. R. 1965. Handbook of Refrigerating Engineering, 2 Vols. Avi Publishing Co., Westport, Conn.

YODER, R. J., and DODGE, B. F. 1952. Heat transfer coefficients of boiling Freon 12. Refrig. Eng. 60, 156–159, 192–195.

Warehouse Construction and Equipment

CHAPTER 3

Birth of Cold Storage and Advent of Small Commercial Cold and Freezer Warehouses

PRELIMINARY CONSIDERATION OF COLD AND FREEZER WAREHOUSE DESIGNS

Cold and freezer warehouse designs may be classified in either very general or more specific types of storage service. As general types they may be divided then as: (1) warehouses in which only a general class of produce is stored; or (2) warehouses in which is stored a specific class of products requiring different temperatures and atmospheric conditions.

The consulting refrigeration engineer may classify cold and freezer storages by relative size and useful volume capacity. Captions that are accepted by the refrigeration trade that are in common use by the commercial industries are: (1) *small* cold and freezer rooms; (2) *intermediate* cold and freezer rooms; and (3) *large* cold and freezer storage rooms.

For more specific naming of warehouses when used for a single or special class of refrigerated storage of products, the rooms bear captions such as: (1) Meat, Poultry and Fish Cold and Freezer Storage Rooms; (2) Pear and Apple Storage Rooms; (3) Tropical Fruits and Vegetable Storage Rooms; (4) Peanut, Walnut and Pecan Cold and Freezer Warehouses; (5) Cold Storage Warehouses for Specially Processed Food; and (6) Fur and Woolen Storage Rooms.

This Manual will include several chapters covering these general and specialized fields of low temperature warehousing.

COLD AND FREEZER WAREHOUSING

Cold and freezer warehousing may be divided into still other temperature classifications, such as:

41

Cold and Chilled Storage

Cold and chilled storage of produce may have several types of use-fulness. This space may be used to hold any perishables that can with-stand storage down to freezing temperature without loss. It may be used as in the case of meats for a period of chilling, or the space may be divided up to handle tropical fruits and vegetables such as bananas, yams or citrus items that keep better in space held at from 40° to 60°F.

TABLE 3.1

TOTAL FREEZER STORAGE HOLDINGS AS OF JANUARY 1, 1968 IN THE UNITED STATES

Frozen Fruits	Lb in Freezer Storage Jan. 1, 1968 000 Omitted
Apples	59,478
Apricots	9,174
Blackberries	25,961
Blueberries	36,024
Boysenberries	12,195
Cherries	49,823
Grapes	8,890
Peaches	48,425
Raspberries, red	24,190
Raspberries, black	2,846
Strawberries	145,627
Other frozen fruits	85,823
Total frozen fruits, lb	508,456
Frozen Juices	
Orange juice	364,275
Other frozen juices	169,419
Total fruit juices, lb	533,694
Frozen Vegetables	
Asparagus	18,726
Beans, Fordhook lima	44,851
Beans, baby lima	78,403
Beans, snap, regular cut	125,239
Beans, snap, french style	49,542
Broccoli	64,344
Brussels sprouts	35,942
Carrots	78,319
Cauliflower	27,779
Corn	231,867
Mixed vegetables	34,543
Peas, green	268,008
Peas and carrots, mixed	17,496
Potatoes, French fries	386,030
Spinach	61,942
Other frozen vegetables	230,159
Total frozen vegetables	1,753,190

Freezer Holding Storage

Freezer holding storage space is generally used for short term storage of frozen products that do not require further freezing. This storage in most warehouses is very active on the so called "in-and-out" list.

TABLE 3.1 (*Continued*)

TOTAL FREEZER STORAGE HOLDINGS AS OF JANUARY 1, 1968 IN THE UNITED STATES

Frozen Poultry	Lb in Freezer Storage Jan 1, 1968 000 Omitted
Broilers or fryers	41,058
Roasters	15,583
Hens, fowls	71,454
Turkeys, whole birds	325,783
Turkeys, other	43,744
Total turkeys	369,527
Ducks	4,383
Unclassified	45,765
Total frozen poultry, lb	9i7,297
Frozen Eggs	
Whites	8,865
Yolks	21,539
Whole or mixed	54,988
Unclassified	1,855
Total frozen eggs, lb	87,247
Frozen Fish[1]	
Blocks and slabs	27,098
Cod	11,628
Flounder	10,494
Haddock	6,850
Halibut	3,362
Ocean perch	17,994
Pollock	870
Salmon	454
Whiting	1,517
Other fillets and steaks	5,813
Fish sticks and portions (raw and cooked)	13,638
Crabs (incl. crab meat)	2,894
Lobster tails (spiny)	3,759
Oyster meats	1,016
Scallops meats	1,272
Shrimp:	
Raw (headless, shell-on)	41,220
All other (incl. breaded)	21,224
Squid	1,396
Other shellfish	1,367
Total frozen fish and shellfish	173,866

[1] Source: U.S. Dept. Interior, Bur. Comm. Fisheries.

Sharp Freezing Holding Storage

In any active frozen food area this storage room is usually very active since it functions as both the freezer room (except other than for quick and instant frozen food freezing) or it may function as holding storage for items that should be kept at temperatures between $-20°$ to $+20°F$.

Very Sharp Freezer Holding Storage

This storage room or warehouse should be a long term holding storage room space for products that have been frozen at very low temperatures and are to be held for some months in a completely frozen state. This type of storage has not been in great use over the past few years. Inadequate refrigeration for low temperatures ($-20°F$), and high demand for holding storage rooms in public warehouses have virtually eliminated this type of storage. It is found, currently, in some specialized storages and will no doubt become more in evidence as new warehouses are built, especially with increased cryogenic refrigerant storage capacity.

STORAGE CLASSIFICATION BY LOCATION

Another classification of cold and freezer storage that is commercially extant is described as: (1) storage-on-location; and (2) storage in transit. On-location storage permits immediate cold and freezer storage as the produce comes from the field, abbatoir or processing facilities. This reduces the time of possible spoilage or deterioration and assures more positive preservation as chilled or frozen. Storing-in-transit offers the following advantages: (1) the transportation rate to a distant consumer center can be adjusted to a rate not much greater than a single transport rate from the originating warehouse to the ultimate delivery location; (2) produce can be held at freezer storage junction centers from which quick delivery can be made with a wide variety of choice to where it can be delivered; and (3) the junction storage centers of the warehouse in transit can be geared to large holdings of featured products at a storage rate less than those of most urban center warehouses where real estate costs, taxation, delivery space and air purity conditions are more satisfactory. A more recent classification is made necessary by the development of cave and quarry refrigerated warehousing. Since many of the design problems related to undersurface cold and freezer warehousing are much different both in theory and practice than those above ground, a complementary chapter is given on *Underground Cold and Freezer Storages* (Chap. 5).

Current underground enterprises are small in number but in average tonnage handled per installation, they far exceed those of the established surface type refrigerated warehouses of North America.

The principal factors that must be controlled in all perishable foods warehousing are: temperature, relative humidity, air movement over the product, and odor control. The dry bulb temperature may be taken either on a thermometer or a pyrometer, the relative humidity and related humidity values are determined by the depression of temperature as indicated between the bare dry thermometer bulb and the wetted thermometer when air passes over them; this is named "wet bulb depression." Mechanical fans are a part of any cold or freezer storage system since uniformity of temperature conditions in all parts of the room is required. Odor control is essential if freshness of food quality is to be maintained.

EARLIER TYPES OF COLD STORAGE

Before the beginning of the Christian era crude methods prevailed where snow and ice were used to utilize these products of cold to, not only cool their drinks but also to chill their fish and available meats, especially to extend the season of wintry temperature in natural and excavated caves and underground excavations for preservation of foodstuffs in ice and snow cooled chambers. In the arctic regions this extended ice preservation season could be more readily accomplished by storing in massive above ground igloos or ice rooms. The early slaughter houses of Rome were front street industries. The incoming cattle and sheep being delivered for slaughter delayed traffic, and the offal obstructed sewerage. The odors are recorded as most unbearable. By imperial order, all slaughtering was subsequently moved into one large center of meat and fish processing with some provisions for snow and ice cooling for the royalty and the very important citizens of the city. Even the sanitary methods of using the available commercial ice were proclaimed by imperial order, this to prevent the ravages of typhus and typhoid fevers caused by the marketing of ice and snow from polluted lakes and streams.

In France, until the period of Napoleon, the storage of meat by ice refrigeration was limited to serving the uppercrust of the citizenry only. This led to an early introduction of mechanical refrigeration in Paris restaurants and hospitals. Within the United States ice cooled meat and fish storage was confined largely to the northern states where extensive ice houses extended from Boston to Seattle. While generally not well-organized, most communities of two thousand inhabitants or more had lake or river ice storage houses to serve the community with refrigeration for food and liquids.

Mechanically Refrigerated Cold Storage

In the United States Benjamin M. Nyce, who started out in Indiana, built his first storage units for storing Indiana apples using ice cooling with circulating pans. He took out his first patent US No. 21,977 in 1858 on a cold storage house design. Nyce continued to expand his building program introducing machine refrigeration cold storage houses from Cincinnati to New York City. The first refrigerated frozen meat processing and storage must be credited to Dr. Henry Peyton Howard of San Antonio, Texas. In 1868, Dr. Howard constructed a 25 by 50 ft cold storage room in New Orleans and shipped sides of frozen Texas beef to this New Orleans room. This beef was sold to hotels, hospitals and charitable institutions of the Louisiana city. He completed his experiment by presenting a banquet of beef steaks to the important meat buyers of New Orleans. The New Orleans Picayune, July 13, 1869 reported that this Dr. Howard's experiment was a great success. Henceforth even in the semi-tropical areas of North America, the freezer storage of all grades of beef and mutton was a most acceptable product. Storage houses in the southern states of the United States that depended for cooling on ice or salt-ice were virtually unknown except in the mountainous regions where ice and snow were available several months of the year. The retail price of northern lake ice delivered by steamship to southern cities was marketed at 5 to 10 cents per pound. At this same period, first grade beefsteak and hams sold in the city meat markets for 2 to 3 cents per pound. In some Texas cities, the barter price of fresh steak was 5 lb of this first grade meat for 1 lb of delivered lake ice. The most economical method for human beings to preserve their perishable meats and fish in the Gulf States prior to 1870 was to smoke, pickle or dry all surpluses except the more delicate portions that did not lend themselves to such processing.

To most Americans of Western European and British heritage, meat and fish that had been frozen were not desirable for human consumption. Northern United States and Canada, however, took the leadership on freezing fish on the surface of frozen lakes, and packing their freshly killed pork, beef and mutton in available winter ice and compressed these in snow, sometimes with salt added. To lower the temperature this freezing operation was usually done in large boxes or barrels in unheated rooms at the season when the thermometer dropped to $-15°F$.

These containers were often covered with oat straw or timothy hay that could maintain the low temperature from December to May if properly protected. Canadians and Americans residing in northern states were the first to accept frozen beef as equal to, or superior to meat chilled and cured at 40°F. This was 3 or 4 decades before the Australians were able to convince their Western European meat buyers that frozen beef and mutton were edible and even superior to "cured" domestic product that usually had a moldy taint within 3 weeks cold storage.

SMALL COMMERCIAL COLD AND FREEZER WAREHOUSES

INTRODUCTION

Cold storage at 32°F or above, prevailed until the beginning of the twentieth century. Even as late as 1930, more than 80% of the cold and freezer storage capacity in America was devoted to above freezing temperatures. Ice and salt were used in the small cold storage rooms and dairy product plants. The principal demand for room temperatures as low as 0°F was for ice cream hardening and butter storage. Hundreds of ice storage houses served as vegetable and fruit emergency cold storage facilities. The locker storage industry of today spent its infancy in the ice house of the first quarter of the twentieth century.

The early freezer storage rooms were operated as a community service and usually at a financial loss until 1925 when quick frozen foods created a demand for 0°F storage. Almost simultaneously the demand for locker storage facilities in the rural areas of North America by farmers and ranchers made small freezing and storing plants a necessity. While the Great Depression reduced the investments, many regions could conveniently make the economic pressure of the producers of meats, fruits and vegetables for a cold storage facility in which they could keep their perishable produce.

COMBINATION ICE PLANTS AND COLD AND FREEZER WAREHOUSES

Many of the cold storage plants of the period prior to World War I were associated with ice plants. The refrigeration facilities and operating personnel for producing the ice and for refrigerating the cold storage plants were the same except that the cold and freezer storage division and the ice plant usually had separate business managers.

Small Freezer Storage Rooms

The average freezer storage room prior to 1925 had a storage capacity of less than 50 tons of refrigerated area. The principal products to be found in these rooms or vaults were tub butter, frozen eggs or ice cream. The latter was on a holding room status of less than ten days. Even frozen eggs and butter were on short term storage for local merchants, bakeries and factories. When the freezer storage rooms exceeded their capacities, the butter and eggs were transferred by freezer rail to larger centers where freezer storage was featured, such as rail transfer centers.

ESTIMATING FOR FREEZER STORAGE REFRIGERATION IN SMALL INSTALLATIONS

For estimating purposes, each 12 cu ft or 1 lb of air will require $\frac{1}{4}°$ of freezer room temperature, cooling or heating. The specific heat of air is approximately 0.24.

Every pound of air would require 0.24 Btu to cool or heat it 1°F. One Btu would then heat or cool approximately 50 cu ft of air 1°F. For a freezer room at 0°F and outside air at 100°F 2 Btu are required to cool 1 cu ft of air from 100° to 0°F.

The refrigeration required, other than for the product heat, in a freezer storage room is heat entering by conduction through the wall, the floor and the ceiling and the air leakage infiltration coming in by convection mostly by the entrance doors, elevator and shafts, etc.

For walls with 5 in. of cork, the conduction can be considered as 0.05 Btu per sq ft per hr per degree temperature difference.

Computation

For a temperature difference of 100°F and 0.05 Btu conduction factor, the conduction per sq ft would be 5 Btu per hr.

For 1,000 sq ft of wall at 100° F for a 0° room, it will equal 5,000 Btu of refrigeration.

5,000 Btu would equal $\dfrac{5,000}{12,000} = 0.416$ tons of refrigeration.

Then: For estimating purposes use 0.4 tons of refrigeration per 1,000 ft of wall surface.

Or: One ton of refrigeration will take care of 2,500 sq ft of wall space.

Floors.—In a freezer room at 0°F and an outside temperature of 100°F the floor heat gain will be related to the mean ground temperature at that latitude. This mean ground temperature would be 70°F at Austin, Texas; 60°F at St. Louis, Missouri, and 50°F at Milwaukee, Wisconsin.

   1 ton of refrigeration per 4,800 sq ft of floor in Milwaukee
   1 ton of refrigeration per 4,000 sq ft of floor in St. Louis
   1 ton of refrigeration per 3,400 sq ft of floor in Austin

Ceiling.—If the freezer room is an insulated single floor with 10 in. of cork in the ceiling, the total conduction factor will be about 0.025 if there is a good built-up roof over it and the ceiling is plastered. In summer a black top roof may equal 180°F in sunlight. If the freezer is held at 0°F and is a single floor building with a black top roof, the differential temperature will then be 180°F. There will be 0.025×180, or 4.5 Btu per hr heat conduction per sq ft of ceiling or for 1,000 sq ft there will be 4,500 Btu per hr. This would be $\dfrac{4,500}{12,000}$ or ⅜ ton of refrigeration per 1,000 sq. ft.

CONDENSER SYSTEMS FOR COLD STORAGE ROOMS

Estimating Additional Power Required for Cold Storage Refrigeration Systems

Using Air Cooled Condensers for Installations Under 50 Tons.—Additional power requirements for air cooled condenser fan and increased compressor load are 25% greater for air cooled systems over water cooled. For water cooled systems, the water pump power load will average a 50% offset. This nets an overall 12.5% power load extra, using air.

First cost of water cooled versus air cooled system will be 25% greater for water.

The depreciation rates: water system—10-yr life; air system—20-yr life. For permanent installations the correct answer may be in favor of the air condenser system even up to 100 tons refrigeration load.

The figures shown above are based on the use of one of the so called low pressure refrigerants such as R. -12, R. -22 or R. -502. Ammonia refrigeration is generally used only above ratings of 100 tons.

BIBLIOGRAPHY

ADLER, C. 1967. Submersibles seek untapped supplies of fish for tomorrow's freezers. Quick Frozen Foods 29, No. 9, 139–144.
AMERICAN MEAT INSTITUTE FOUNDATION. 1960. The Science of Meat and Meat Products. W. H. Freeman Co., San Francisco.

ANDERSON, O. E., JR. 1953. Refrigeration in America—A History of a New Technology and Its Impact. Princeton Univ. Press, Princeton, N.J.

ANON. 1891. The frigerized meat export trade. Ice and Refrig. *1*, 88-89.

ANON. 1946. Frozen food industries—plant layout, cost of processing, marketing. Univ. Ark. Bur. Res. Inform. Ser. *1* Rev.

ANON. 1954A. The Birdseye story. Quick Frozen Foods *17*, No. 2, 55–61, 63, 65, 67, 69–70, 73, 75.

ANON. 1954B. An industry attains maturity. Quick Frozen Foods *17*, No. 2, 77, 79, 81, 83, 85, 87, 91, 93, 95, 98–99, 101–103, 105–112.

ANON. 1955A. Frozen vegetables priced under canned in survey. Quick Frozen Foods *17*, No. 9, 88–89.

ANON. 1955B. Convenience and quality factors cause housewives to buy more frozen juices. Western Canner Packer *47*, No. 4, 39–40.

ANON. 1955C. Cold cash. Arthur D. Little, Inc. Ind. Bull. *323*, 2–3.

ANON. 1955D. Trends in our eating habits. U.S. Dept. Agr., Marketing Serv. Natl. Food Situation, NFS *73*.

ANON. 1966A. Frozen sea food value soared 26.8% to within sight of a billion dollars. Quick Frozen Foods *29*, No. 3, 125–128.

ANON. 1966B. Warehousing and transportation report. Quick Frozen Foods *29*, No. 3, 151.

ANON. 1966C. 79 Refrigerated warehouses built in two-year period 1963–1965. Quick Frozen Foods *29*, No. 2, 99–101.

ANON. 1966D. Freezer storage space capacity USA. Quick Frozen Foods *29*, No. 4, 100.

ANON. 1966E. Gross refrigerated storage capacity. Quick Frozen Foods *29*, No. 4, 99.

ANON. 1966–1967A. Processors and products—fruits, juice concentrates, vegetables, fish, meat, poultry, prepared foods. 19th Ann. Edition, Sect. 1, 17–161. Directory Frozen Food Processors; Quick Frozen Foods, E. W. Williams Publications, New York.

ANON. 1966–1967B. Refrigerated Warehouses. 19th Ann. Edition, Sect. 6, 344–374. Directory Frozen Food Processors, Quick Frozen Foods, E. W. Williams Publications, New York.

ASHRAE GUIDE AND DATA BOOK. 1967. Sect. 2 and 7. Am. Soc. Heating, Refrig. Air Cond. Engr. New York.

BANKS, M. R. 1954. Capacity of refrigerated warehouses in the United States (as of Oct. 1, 1953). U.S. Dept. Agr., Agr. Marketing Serv., Statistical Bull. *148*.

BESNOR, D. B. 1891. Practical Cold Storage. The handling of poultry. The best way to freeze and pack dressed poultry, etc. Ice and Refrig. *1*, 279–280.

BIRDSEYE, C. 1953. Looking backward at frozen foods. Refrig. Engr. *31*, 1182–1183, 1250, 1252, 1254.

BITTING, H. W. 1955. FF use in preserves, pies and ice cream. Quick Frozen Foods *17*, No. 10, 129–131.

CARLTON, H. 1946. The freezer locker plant is going commercial. Food Inds. *18*, 1542–1544, 1672, 1674, 1676.

DAVIS, L. L., and RODGERS, P. D. 1948. Planning a frozen food business. Virginia Agr. Expt. Sta. Bull. *419*.

DIEHL, H. C., and HAVIGHORST, C. R. 1945. Progress and prospects of the frozen food industry. Food Inds. *17*, 261–278.

EASTWOOD, R. A., and SCAALON, J. J. 1952. Operating cost of 15 cooperative poultry dressing plants. U.S. Dept. Agr., Farm Credit Admin. Bull. *70*.

FITZGERALD, G. A. 1956–1957. Fruits and vegetables. *In* Air Conditioning Refrig. Data Book, 6th Edition. Am. Soc. Refrig. Engr., New York.

FOX, J. 1955. Spectacular rise of the concentrates emphasized by their tenth anniversary. 1955 Frozen Food Factbook. Natl. Wholesale Frozen Food Distributors Assoc., New York.

FRANKLIN, H. L., and Martin, S. 1967. FF per capita consumption rose to 59,389 lbs in 1965. Quick Frozen Foods *29*, No. 7, 37–43, 154–155.

HARDENBEECH, W. 1955. Four factors shaping frozen meat future. Quick Frozen Foods *10*, No. 5, 126–127.

HASTINGS, W. H., and BUTLER, C. 1956–1957. Fishery Products. Air Conditioning, Refrig. Data Book. Am. Soc. Refrig. Engr., New York.

HAVIGHORST, C. R. 1944. What's ahead for frozen foods. Food Inds. *16*, 435–439.

HAVIGHORST, C. R. 1955. So you are going into freezing. Food Inds. *17*, 1471–1475.

HAVIGHORST, C. R., and DIEHL, H. C. 1947. Frozen food Rept. 2, Part 1, Transportation, warehousing, frozen food distributional-facilities map. Food Inds. *19*, 3–26; Part II, Marketing. Ibid. 148–160.

KAUFMAN, V. F. 1951. Costs and methods for pie-stock apples. Food Eng. *23*, No. 12, 97–105.

MANN, L. B., and WILKINS, P. C. 1953. Merchandising commercial frozen foods by locker plants, 1952. U.S. Dept. Agr., Farm Credit Admin. Misc. Rept. *1*, 175.

MAYHEW, E. E. 1952. Cost accounting for the frozen food packer. Quick Frozen Foods *14*, No. 8, 95–98, 326, 328–329.

PAULUS, R. C. 1955. Northwest berry industry trends. Western Canner Packer *47*, No. 6, 75–79.

PENNINGTON, M. E. 1941. Fifty years of refrigeration in our industry. U.S. Egg and Poultry Mag. *47*, 554–556, 566, 568, 570–571.

RADIT, W. H., and HAMER, A. A. 1961. Protection of rail shipments of fruit and vegetables. U.S. Dept. Agr., Agr. Handbook *195*.

SHEAR, S. W. 1955. Trends in United States fruit production and utilization. Western Canner Packer *47*, No. 6, 70–73.

STAPH, H. E. 1949. Specific heat of foodstuffs. Refrig. Eng. *87*, 767–771.

STEIN, A. 1945. Rise of the frozen food industry. Conference Board Industry Record *4*, No. 10, 1–6.

STEVENS, A. E. 1955. Outlet for concentrates. Quick Frozen Foods *17*, No. 9, 139–140, 204, 206.

STEVENSON, C. H. 1899. The preservation of fishery products for food. U.S. Fish. Comm. Rept. *1898*, 335–363.

TEWSBURY, R. B. 1953. Rail transportation of perishable foodstuffs. Refrig. Eng. *61*, 52–54, 108.

TRIGGS, C. W. 1955. Trends in production, processing, and distribution. Fishing Gaz. *72*, No. 6, 83–84.

VANDERVAART, S. S. 1916. Commercial uses of refrigerating machinery. Ice and Refrig. *51*, 179–186.

WENZEL, F. W., MOORS, E. L., and ATKINS, C. D. 1952. Factors affecting the cost of frozen orange concentrate. Quick Frozen Foods *14*, No. 8, 101–102.

WHITMAN, J. M. 1957. Freezing points of fruits, vegetables and florist stocks. Marketing Res. Rept. *196*, U.S. Dept. Agr., Washington, D.C.

WILLIAMS, E. W. 1954. Frozen foods 2000 A.D.—a fantasy of the future. Quick Frozen Foods *16*, No. 7, 101–108.

WILLIAMS, E. W. 1955A. In which direction is the industry going? Quick Frozen Foods *17*, No. 8, 97–102.

WILLIAMS, E. W. 1955B. Should a wholesaler go into frozen foods? Quick Frozen Foods *17*, No. 10, 49–51, 171–172.

WOOLRICH, W. R. *et al.* 1933. The Latent Heat of Foodstuffs. Tenn. Engr. Expt. Sta. Bull. *11*, Knoxville, Tenn.

CHAPTER 4

Large and Intermediate-Sized Cold and Freezer Storages

HISTORICAL

Prior to World War II most cold storage space consisted of coolers at above freezing temperatures and very little space was devoted to freezers or the freezing of products. It is true that the locker freezer storage plants had become quite popular but these storages were relatively small. The larger commercial cold storage plants were devoted mostly to the storage of cooler products such as eggs, apples and other cooler-held products. Freezer storage was available in rather limited amounts and mainly used for the storage of frozen fish, meats, butter, ice cream and a few other products. Some experimental work was being done on freezing and freezing methods but the actual market impact of frozen foods had not yet taken place.

The advent of World War II saw a tremendous upswing in the demand for both cooler and freezer space. A great many cold storage plants were erected by the government for military use. Large amounts of foods were frozen, creating a demand for freezer space. To make this available, many cold storage plants converted existing cooler rooms to freezer space. This was usually accomplished by the addition of surface to the existing low side equipment in the cooler room and by lowering the evaporator temperature in the low side equipment by the addition of more compressors or diversion of existing compressor equipment. Insulation was sometimes added to the rooms but more often was not. The resultant freezer space was makeshift at best. Quite often the insulation, not being properly sealed and of inadequate thickness for freezing temperatures, would pop off the walls with distressing results to the stored products. Some insulations stayed in place for quite some time and allowed use of the room as a freezer. Although results were not too good with the converted rooms, they did serve the purpose of providing sorely needed freezer space. A good proportion of the converted rooms were ruined by the conversion and had to be rebuilt at a later date. The most serious problems with a converted room were usually encountered when the room was returned to cooler service. As the room defrosted, it quite literally came apart as the ice and frost

53

that had accumulated during its freezer use was about all that held the room together.

After the war ended, quite a boom in cold storage space requirements was experienced. Not only were a number of new plants erected, but the form and concept of cold storage plants changed. Before the war, most cold storage plants were multi-storied buildings located in congested districts. The post war period saw rapidly rising labor costs and the development of efficient materials handling equipment. Battery driven fork-lift trucks were in some use and considerable work was done to equip them for use in freezer rooms. The newer warehouses were being erected as a single story structure with large rooms and designed for palletized operation with fork-lift trucks to speed handling and to permit more economical use of space. High ceilings came into being since fork lifts could effectively use higher piling heights efficiently. Work spaces between rooms and wider docks evolved with more attention to an increasing number of refrigerated over-the-road transport trucks. All of this was done in the interests of faster and more economical handling of merchandise, in and out of storage.

Pre-World War II cold storage plants were refrigerated, for the most part, with refrigerant or brine in pipe coils, the pipe coils being arranged in banks in the various cold rooms. As newer plants evolved, so did the form of the low side equipment used in the storage rooms. The blower type coil, consisting of a compact pipe tube bundle in a

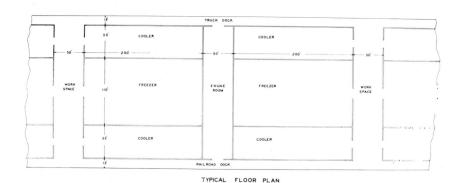

TYPICAL   FLOOR   PLAN
ALFORD   REFRIGERATED   WAREHOUSES

Courtesy of Alford Refrigerated Warehouses, Dallas

FIG. 4.1. TYPICAL FLOOR PLAN FOR LARGE MODERN WAREHOUSE WITH LARGE ROOMS

Note work bays extending through warehouse for ease in working various rooms either to truck dock or railroad dock.

casing with a fan forcing air over the tube bundle and out into the room, came into more or less general use. Still more refrigeration surface was compacted into a small space by the addition of metal fins to the pipes in the coil bank. The new blower coils were small and compact and much easier to install than the bulky pipe coils used in the earlier plants. They required little if any building structure bracing due to their light weight. The first cost was much less than the conventional pipe coil.

Historical Mistakes

Errors were made in the selection of the early blower coils. Pipe coils had usually been selected on the basis of 15°F or more temperature difference between room air temperature and refrigerant temperature. Early selection of blower coils was quite often based on this same selection method. With forced air circulation over the blower coils in a room, humidities were quite low and blower coils soon earned a rather bad reputation for excessive dehydration of foods and stored products. It was the belief of many that any air circulation in a cold room was bad for stored products and would cause excess desiccation of the product. It took some time for the general recognition that the dehydration in storage rooms was not caused primarily by the air circulation, but from excess removal of moisture from the air caused by the high temperature difference between room ambient air temperature and refrigerant temperature. The answer to this was rather obvious and much better results began to appear as temperature differences were kept smaller and it began to be generally realized that the humidity in a room could actually be controlled, within limits, by the amount of coil surface and refrigerant temperature with relation to the room temperatures. By about the year 1950, blower coils were rather widely accepted and used by cold storage installations. A temperature difference of 10°F between room air and refrigerant temperature was more or less accepted as a design standard for room humidities of the order of 80% in cooler rooms. Temperature differences were then adjusted up or down from this to obtain desired room humidities. In cooler rooms, a range of about 65% to 85% RH can be maintained by varying the refrigerant temperature with relation to the room temperature. Above or below this figure, other means will usually be required to supplement the refrigeration to obtain the desired results. High humidities are usually desirable in cooler rooms and also in freezer rooms and a temperature difference greater than 10°F between ambient air and refrigerant temperatures is usually not recommended.

SOME DESIGN FACILITIES OF LARGE COLD
AND FREEZER STORAGE

The modern cold storage room can vary from a few thousand square feet to an acre or so in size. Ceiling heights are usually from 20 to 25 ft. Rooms are most often constructed with steel columns and beams and with insulation as an envelope around the room. Access to the room is normally through a set of double doors in the larger rooms. Electrically operated doors are being used quite commonly as outside doors for freezers and coolers. In freezer rooms, there is quite often a vestibule, or air lock leading into the room. Both ends of the air lock are usually closed by electrically operated or air operated doors. In large freezer rooms, it is not uncommon to find two sets of vestibules and doors; one used for incoming and the other for outgoing loads. Air curtains are also used at times over door openings with varying degrees of success.

Room refrigeration equipment normally consists of fan-coil units in multiple or singly, depending on the size of the room. Fans are propeller or centrifugal depending on whether duct work is used. Quite often, ceiling hung units are used, arranged in a pattern to give good air coverage of the room without the use of duct work. Some designers use floor units with short air ducts up to ceiling level or a complete duct distribution air system. The floor unit lends itself to ready maintenance and repair but takes up valuable floor space and is also subject to damage from materials handling equipment. Duct work should be kept to a minimum. It will usually be found that cold air, particularly in freezer rooms, will form a rather even temperature pattern in the room regardless of how it is distributed in the room. An intricate air distribution system is not normally required in a cold room as good results can be obtained by scattering the refrigeration units in the area without duct work. Various methods of defrost are used in freezer rooms. These methods are discussed in another chapter.

REFRIGERATION CALCULATIONS

There is really no exact method of calculating the heat losses from a cold storage room. Many tables and sample calculations can be found so that losses can be calculated, supposedly to the last Btu, but in the last analysis, it really becomes a matter of judgment. Most designers will start out by calculating the heat losses from a room through the insulation by the use of known factors available for the various types of insulation. To this they will add product load, electrical loads, infiltration and other losses peculiar to each installation. When these calculations are all made, the experienced designer will take a long hard look at his results and then add load to the calculations. This factor added to the calculated load is variously known as a "factor of ignorance" by some and as a "safety factor" by others. It should actually be a judgment factor of experience gained from previous designs and countless observations of existing rooms and the amount of refrigeration they require to maintain proper temperatures.

It would probably be possible to calculate the losses from a cold storage room and design to these losses exactly and obtain the desired room temperatures; if the room could be locked and access to it controlled; if the insulation were perfect; if incoming loads were limited exactly to design; if the outside temperature never exceeded the design temperature and all other circumstances of the room could be exactly controlled. Unfortunately for the designer, this is never possible, nor is there any way to make it possible. When an owner decides to invest in a cold storage facility, he expects to receive a reasonable return on his investment. He cannot be expected to turn down merchandise or work a room at a pre-determined rate when it is possible to work it faster and obtain more revenue from the room. It is not up to the designer to force the owner to operate at the design load; but to design a system that will operate at whatever load the owner chooses to impose on it. Of course, there are some limits that must be put on the design at the outset but these should be kept to a minimum and should be clearly understood between the owner and the design engineer. To avoid misunderstandings, a written summary of design conditions and limitations is quite often desirable. The experienced designer should anticipate loads that may occur and other factors that the owner may not even be aware of.

It is always a good plan, when starting a design for a cold storage facility, to make load calculations of the insulation losses and other losses that are likely to occur during the use of the various rooms. These calculations will give a figure that is a good starting point, or base, for the actual load that may be encountered. There are a number of items that should be considered in addition to the standard load calculations.

Insulation and Vapor Barrier Faults

Most insulating materials will do an effective job when installed with a good vapor barrier. Over the years, however, some deterioration will most assuredly occur. This can be more or less serious depending on the type of insulation used and the care with which it and the vapor seal were installed. Insulation "diseases" are insidious and usually show up very gradually and quite often progress to a rather serious point before being noticed. They start very slowly and sometimes do not start until quite a few years have passed. It is not easy to be sure that insulation is performing the way it should. A very gradual warm up of a room usually observable only over a period of months is a good sign of insulation deterioration. This, coupled with the necessity of more frequent defrosting of the coils in freezers, is also indication of failure of the vapor seal. Repairs are difficult to make since it usually requires access and inspection of spaces difficult to reach. Any load calculation should take into account that there is bound to be some deterioration of the insulation and vapor seal over a period of years no matter how carefully the insulation and seal are erected.

Door Openings.—Cold storage operation becomes more and more hectic as business increases. Doors into cold areas have a tendency to be left open for longer periods. It is difficult to say just how much heat and moisture can be drawn into a cold room through an open door: but anyone who has observed an open freezer door into an area that is warm and humid can appreciate the fact that a great amount of heat is transferred while the door is open and also large quantities of moisture vapor can be drawn into the cold room. No accurate calculations can be made of these losses since conditions are constantly changing and there is no real pattern to follow. These losses are a substantial load, however, and in some instances may be almost the dominant load in a room. Humid air entering the room through an open door can also cause frequent defrosting in freezer rooms. It is sometimes difficult to determine whether frequent defrosts are occasioned by vapor leaks through the insulation vapor barrier or vapor transmission through open doors. Observations of frost during periods of high room use against build up when the room is shut, as on a weekend, will show if excess infiltration of moisture is present.

Product Load.—Some product load is usually figured in the initial calculations of any plant. For safety, this load should be figured near its possible maximum. In freezer storage rooms, it is good practice to conclude that cars of frozen merchandise will be received partially thawed which will require considerable tonnage of refrigeration to refreeze. Also that products from sharp freezers will quite often be transferred to holding rooms before being completely frozen to finish freezing in the holding room. This is not good practice but it is quite frequently resorted to when merchandise to be frozen piles up beyond the capabilities of the plant sharp freezing facilities. There may even be times when loads of unfrozen merchandise will be placed in a hold-

Courtesy of Alford Refrigerated Warehouses, Dallas

FIG. 4.2. DOUBLE VESTIBULE DOORS WITH AIR OPERATED OPENERS USED TO SEAL OPEN-INGS INTO REFRIGERATED SPACES

Doors are glass coated to seal out moisture and improve toughness of surface. Continuous hinge of synthetic material prevents frost buildup and is resistant to hard knocks and breakage.

ing freezer for freezing; sometimes with the use of auxiliary fans and sometimes not. Again, this is not good practice but the good design engineer needs to make provision for all foreseeable conditions, both good and bad.

Outside Conditions.—Most plants are designed using outside weather data gathered over the years and based on actual recorded temper-atures. Consideration needs to be given to that hot spell which may

Courtesy of Alford Refrigerated Warehouses, Dallas

FIG. 4.3. TYPICAL DOCK SCENE ALONG DRY STORAGE DOCK
SHOWING CAR UNLOADING AND MERCHANDISE MOVEMENT
ALONG DOCK

only occur once in 5 or 6 yr where the outside temperatures may be warmer than accepted design conditions for a rather extended period.

There are many tables published giving tonnages for storage rooms. Most of them are based on theory and practice and are of some value. Because of the many variables involved in calculating room losses, it is impossible to produce a table that will meet all conditions. After use of any table, it is well to look at the design and add in for some of the variables to assure adequate tonnage. It is always well to keep in mind that it is extremely difficult to put too much low side equipment in a room. Oversized equipment can overcome a multitude of design deficiencies.

Design by Experience

From the foregoing paragraphs, it can be seen that design is quite often primarily a matter of experience. It should always be remembered that commercial cold storage plants are built to make a reasonable return to the owner on his investment. When designs are inadequate and rooms cannot be used to their full potential and beyond, then the owner is not getting his full return. A plant cannot afford to turn away business that could have been handled profitably if just a little more equipment had been installed at the start. It should be the aim of the competent design engineer to design for the ultimate load that can be profitably handled. This makes for a good plant and a satisfied owner.

BIBLIOGRAPHY

ASHRAE. 1969. Guide and Data Book, Equipment. Am. Soc. Heating Refrig. Air Conditioning, New York.

ASRE. 1956–1957. Air Conditioning and Refrigeration Data Book, 6th Edition, Applications. Am. Soc. Refrig. Eng., New York.

COOLING, L. F., and WARD, W. H. 1950. Damage to cold stores due to frost-heaving. Ice and Refrig. *119*, No. 1, 41–44.

CHRISTENSEN, P. B., and HAYNES, D. O. Materials handling in multistory cold storage warehouses. Refrig. Eng. *54*, 539–543, 583.

HALLOWELL, E. R. 1949A. The Alford refrigerated warehouses. Refrig. Eng. *57*, No. 3, 226–237, 272–275.

HALLOWELL, E. R. 1949B. Recent developments in refrigeration warehouses. Ice and Refrig. *116*, No. 3, 53–56.

STENCEL, R. A. 1947. The effect of frozen foods on cold storage warehouses. Can. Refrig. J. *13*, No. 10, 15–16.

TRESSLER, D. K., VAN ARSDEL, W. B., and COPLEY, M. J. 1968. The Freezing Preservation of Foods, 4th Edition, Vol. 1. Avi Publishing Co., Westport, Conn.

WOOLRICH, W. R. 1966A. Freezer storage. *In* Handbook of Refrigerating Engineering, Vol. 2. Avi Publishing Co., Westport, Conn.

WOOLRICH, W. R. 1966B. Cold storage, food preservation and fur storage. *In* Handbook of Refrigerating Engineering, Vol. 2. Avi Publishing Co., Westport, Conn.

CHAPTER 5

Underground Cold and Freezer Storages

INTRODUCTION

Underground cold and freezer storage is as ancient as reasoning Mankind. Wherever mountain caverns existed or where ice was found in nature during a few months of each year, caves, man-made or natural, were utilized to some degree of proficiency for preserving food by cold. Where the annual mean temperature beneath the earth's surface fell below 32°F, artificial or mechanical refrigeration was not necessary but such places on the earth's surface were usually thinly populated. These isolated caverns used for perishable food storage were most generally utilized by a few tribal or family associates and were not for public participation. The most notable recent progress in underground freezer storage in North America for large perishable foodstuff tonnages has been that in Eastern Kansas, primarily in the Kansas City area, which accomplished real meritorious advancement subsequent to World War II.

The initial commercial cold and freezer storage to be established underground dates to World War II in the vicinity of Atchison, Kansas.

From the refrigeration files of that period, and especially the magazine, Ice and Refrigeration, of Chicago, Illinois it is recorded that Lt. Col. Ralph M. Ohmstead, Deputy Director of the War Food Administration Office of Distribution, conceived the idea of utilizing the limestone underground quarry chambers as refrigerated storage vaults for the millions of pounds of perishable products to be distributed across the United States under the War Food Administration.

The Atchison quarries had about 300 underground "rooms" or bays that averaged 65 ft square and 13½ ft high that were well-drained and that maintained an average temperature of 55°F.

Historically, caverns and mine rooms in France, Pennsylvania and Minnesota had been employed for Roquefort cheese aging and subsequent storage. They had been used to store successfully this type of cheese for as much as 3 or 4 months at temperatures near 47°F. Ohmstead was also familiar with the western desert ice caverns that the pioneers had used to store food especially during the gold seeking migrations.

The estimated space available in the Atchison limestone mine was

12,000,000 cu ft of piling space. This capacity would accommodate 3,000 carloads of foodstuffs for either dry or highly humid storage as required by the wide range of perishable food products. This original quarry limestone mine had a relatively level floor which the Government proceeded to give a concrete leveling surface. When completed there were 521,000 sq ft of gross floor area of which 60% could be used for net storage. The initial Atchison warehouse equipped for refrigerated storage had three 250-ton ammonia compressors with the essential complementary condensers, brine coolers, evaporators, controls and circulating ventilating fans.

For insulation it was estimated that 10 ft of dried out limestone would be the equivalent of 6 in. of cork. The original problem was to lower the temperatures and the moisture content of the great mass of stone in the overburden, the walls and the floors. In this necessary pulling down operation all 750 tons of refrigeration could be utilized. After 4 or 5 months of cooling and dehumidifying the ceilings, walls and included space, one of the 250-ton ammonia compressors could maintain the desired temperature and humidities.

Immediately after World War II quarrying was initiated by Lorin Quarries Company at Loring, Kansas to provide ballast for the Union Pacific tracks. Later, great quantities of the crushed rock were used as aggregate in the construction of the Kansas Turnpike and for general building construction in the Kansas City area. Quarrying at first was done by open or strip mining, but the unusable overburden became so thick that it was not economical to continue open mining, and thus tunneling operations were started.

By 1952, further mining had progressed some distance into the bluff. A railroad spur that would hold 12 cars was constructed into the mined area. Concrete docks and floors were poured to improve product movement and concrete block walls were built between supporting rock pillars to form rooms. Brine ammonia refrigeration equipment was installed to make cooler rooms and two rooms were set aside to be maintained at 0°F. About 1956, a second spur track holding 24 cars was brought into the area, additional cooler rooms, a second engine room, and one new freezer storage room were developed. The total developed refrigerated area at completion of this work by Natural Storage, Inc. was approximately 750,000 sq ft.

In 1961, Southeastern Public Service Company of Kansas City acquired the property and changed the name to Mid-Continent Underground Storage and added freezer space to bring the total developed area to approximately 1,000,000 sq ft.

The Inland Cold Storage Company was set up about 1954 and began developing refrigerated space in their quarry at 6500 Inland Drive, Kansas City, Kansas. The evacuated rooms and connecting tunnels were high enough above any possible flood water to insure gravity drainage. While internally there existed a few roof faults that provided nuisance seasonal drips, and at some sections running springs prevailed, these presented no problems other than what natural drainage could solve. The overburden consisted mostly of several feet of horizontal rock strata capped with an accumulation of pasture soils of humus, clay and sand of varying depths.

DESIGN AND PERFORMANCE FACTORS TO BE CONSIDERED IN UNDERGROUND COLD AND FREEZER STORAGE

Mr. John G. Mueller, Vice President of Engineering of the Southeastern Public Service Company reports that "from my personal experience, when we developed the freezer space in 1961 we found that it took a little over 30 days to pull the moisture out of the rock before we were able to bring room temperature down to a point where frost would form on coils and defrosting became necessary. For 30 days our 23 ten-ton blower units were pulling moisture out of the air and rock at a rate of nearly 100 gal per day, each. After temperature dropped to frost developing conditions, each unit continued to pull out around 50 gal of water per day. Three years later one 115-ton booster compressor and three of the 10-ton blower units were taken out and transferred to another plant. The room now is being held with 1 ton refrigeration per 1,500 to 1,700 sq ft of refrigeration area. Today we have installed approximately 800 tons of second stage equipment in the entire plant; however, our normal operation requires only a little over 500 tons."

Different mechanical compression systems have been tried in these underground plants, including an ammonia 2-stage brine cooling unit chilling air for the freezer rooms with both rotary and reciprocating compressors; 2-stage air cooled R. 12 systems with direct expansion freezer room coils; single stage air cooled R. 12 compressors with direct expansion freezer room coils and 3 stage centrifugal compressors with mechanical pumps for recirculation to the freezer room coils.

Mr. James L. Williams reports by letter to the authors:

"Ten years of experience with various types of refrigeration systems has proved that an ammonia-brine system provides the best type of refrigeration system to be used in underground freezer storage rooms. This type of system is no more expensive to install than any of the

Courtesy of Southeastern Public Service Co.

FIG. 5.1. UNDERGROUND STORAGE FACILITY

Note ceiling and floor treatment to obtain low humidity required for refrigerating candy.

other systems mentioned and has proved to be far more reliable from an operating standpoint. Maintenance costs on an ammonia-brine system are far less than any system installed thus far which has used Freon as a refrigerant. So much of the design work differs so radically from that which would apply to conventional above ground installations that many of the old and accepted rules of thumb have to be discarded. No insulation is used on the floor or the ceiling and after a period of years the freezing point has penetrated back into the rock an approximate distance of 28 ft. During the pull down to a subzero temperature the ceiling height is reduced up to 1 in. Whether or not the floor rises or the ceiling comes down is something I don't know, but we know that expansion joints should be provided in the floor around all the pillars as well as around the rock walls making up the periphery of the room. Various roadways that have been built inside these underground warehouses to provide ingress and egress for trucks and the various inside truck loading docks have been built for loading and unloading trucks. Railroad spurs have been run into these caves for food express cars. They accommodate many railway cars at both Inland and Mid-Continent underground warehouses."

The excavated limestone strata in both installations is surprisingly horizontal in formation and the removed portion was usually 14 to 16 ft in height. Internally these tunnels and excavated rooms were supported by the original rock pillars about 25 by 25 ft in cross section. The rooms were of variable sizes, usually above 40 by 40 ft. These rooms had been excavated per instructions of the State and Federal mining authorities to assure adequate safety against cave-ins to the excavators.

This region of Kansas is subject to widespread river flooding. Fortunately, the excavated rock strata in both installations is surprisingly horizontal in formation and the removed portion was usually 14 to 16 ft in height. The overburden consists of other horizontal rock strata of several feet in thickness and carrying centuries of accumulation of pasture soils of varying depths of humus, clay and sand.

The excavated rock strata were high enough above the river flood water and internally they had a few roof faults and water springs. The rooms, by nature of the time that had elapsed since excavation ceased, were filled, however, with damp air caused by the annual thermal variations of the ambient air in the many tunnels. This moist air had penetrated some feet into the exposed rock strata. The ceiling did require considerable bolstering-up with $7/8$ in. support rods extended up into the ceiling rock strata for several feet with $1\frac{1}{4}$ in. expansion shields.

In commissioning an old quarry for conversion to underground storage space the factors that must be considered are: (a) the mean ground temperature for that region; (b) drainage and seepage possibilities; (c) mean specific heat and mass; (d) the insulating and heat conductivity value of the earth mass envelope and its heat conductivity values; and (e) the floor and ceiling condition throughout the entire excavation.

The mean ground temperature in any area at and below 20 ft surface level will correspond within 2° or 3°F to ground water temperature of that location. A table of established deep well ground water temperatures for capital cities of the United States is given herewith.

Generally they are a function of the latitude for areas not too greatly affected by either thermal stream waters or a prevailing high altitude cool climate which might raise or lower the mean average regional temperature, respectively.

These ground water temperatures reflect the mean annual temperature of the area under consideration and if the mechanical engineer desires to obtain the mean water temperature for most American cities he can be reasonably sure that it will not vary more than 2° to 3°F from the mean air temperature of that city on the annual basis.

Underground Freezer Storage Floors

Since the normal ground temperature of an underground cavern will approach that of the mean underground water temperature of any specific location, this would be:

°F
45 for Minneapolis, Minn.
56 for Kansas City, Kansas
65 for Dallas, Texas
70 for Austin, Texas
59 for Nashville, Tenn.
55 for Columbus, Ohio
53 for Lincoln, Nebraska
45 for Augusta, Maine
52 for Salem, Oregon

While some refrigeration might be saved by insulating the room floors, the high cost of a serviceable floor surfacing installation together with the inherent problems of floor instability would not justify the investment. The extra power cost of maintaining the desired freezer room temperatures without auxiliary insulation for the floors is much less than the interest, maintenance and depreciation charges for an insulated underground floor. Fortunately, most all man-made caves and surface level mines have maintained a floor level satisfactory for haulage when resurfaced. Some mine stratas dip and make the abandoned mine floor less satisfactory for warehousing or other use. With man-made caves and worked-out mine entrances above ground levels, drainage of inside water is not a serious problem but may be of some aid in maintaining a high humidity where such is desired for some storage items.

Underground Freezer Storage Ceilings

What has been stated about underground freezer floors applies in part to the ceilings, including the freezer rooms. Underground ceilings generally remain without insulation. The great mass of the rock and overburden above each room provides a vast reservoir of intense cold and in itself reduces the heat flow into the refrigerated space.

Structurally, insulating the ceilings would be a much greater and more expensive task than insulating the same area of floor. The ceilings, as they are left after the previous excavators, are much more uneven than the floors. Many rooms contain some moisture that becomes frozen when the cave is refrigerated. If ceiling insulation were attempted, the covering would interfere with the essential inspection of the overhead strata, and also the prevailing lack of insulation does allow free sublimation of any ice that is formed on the room refrigeration coils. It must be recognized that the great mass of overburden in each cold room will be subject to some heaving as the moisture in the mass causes pressure because of the expansion during freezing.

It must be anticipated that with structural changes within the earth itself, primarily due to frost heaving, provision should be made in the installation of doors, for adequate adjustment of door frame shifting as these changes take place.

The complex of doors and insulated walls of the underground storage includes the main entry doors, large and sturdy enough to give full entry height to the rail or motor cars and wide enough for egress and ingress of the food express vehicles without interference. These are primarily security and air flow control doors and need not be so generously insulated.

The room doors requirements are not unlike those of above-ground cold and freezer storage installation. The internal performance of an underground cold storage will call for the same precautions of ingress and egress as from the loading and unloading docks of above-ground storage.

Offices, lunchrooms and warm storage places will require heating for both humidity and temperature control and should be adequately provided. Designers of underground freezer storage facilities should be aware that there is virtually no connection in the natural cooling phenomenon of the ice caves of the mountainous regions that exist even in some warm climates and which gave the early pioneers cold storage facilities throughout the hot season. Most every one of these "ice caves" had a physical condition peculiar to them. All are situated where ice freezing temperatures are experienced in winter and most are located in black basalt rock that is highly porous and permits free air circulation, especially in the winter months. To execute this winter circulation further, most of such caves have a near horizontal entrance to the South and the warmed air at the entrance is displaced by the heavier cold air within. If the winter is severe, the air inside may be at a low enough temperature to continue the freezing process until well in the summer depending upon how much water has dripped in from the higher elevation without the cavern. However, seldom does this ice last until fall, then the cycle is repeated.

Caves excavated for construction stone in lime rock or coal mines with fully excavated rooms on a mountainside may be commercially feasible developments for freezer and cold storage. These installations would use commercial refrigeration compressors and related facilities as the source of mechanical cold. As a workable refrigeration storage structure, the main entrance to such caves or rooms should be level ground and inside, the roadway should be inclined to the main outside opening entrance for assured continuous drainage. The natural structure should preferably be horizontal strata of limestone or dense slate, and all roof strata should be free of appreciable water leakage or pockets. Since the ceiling and floor will eventually attain temperature below $0°F$, there will be some frost heaving in both instances to be anticipated. Each cavernous cold storage room should receive adequate engineering inspection to assure the designers that the floors are adequately drained and the ceiling well supported by tension rods to support any imbalance of stresses with temperature changes and locally provided excessive vibrations.

BIBLIOGRAPHY

ANON. 1951. Another cold storage warehouse to be installed in limestone mine. Ice and Refrig. *121*, No. 1, 32.

ANON. 1952. Refrigerated storage in limestone cave. Ice and Refrig. *123*, No. 2, 38.

ANON. 1965. Inland underground facilities expands, reports record revenue and profits. Quick Frozen Foods 27, No. 6, 116.

INSTITUTE OF FOOD TECHNOLOGISTS. 1965. 25th Ann. Meeting Kansas City, Mo. Visit to Inland Cold Storage Co., Div. Inland Underground Facilities, Kansas City, Kans.

MUELLER, J. G. 1966. Midcontinent underground storage facilities for freezer, cooler and dry storage at Bonner Springs, Kans. Personal Communication to W. R. Woolrich, Sept. 13.

STANLEY, G. E., JR. 1951. Underground cold storage warehouse opens near Kansas City. Ice and Refrig. *121*, No. 2, 45–56.

CHAPTER 6

Insulation for, and Heat Transfer through Cold and Freezer Storage Walls and Ceilings

INTRODUCTION

Heat transfer is a very complex process. Heat may be transferred from one point to another by three processes: conductivity, radiation, and convection. For this section the discussion will be directed to insulation to increase the resistance to heat conduction. In another chapter the conduction of heat and cold for refrigerant condensers, evaporators and heaters is treated. In the case of these heat exchanger units, the interest is in having a high rate of heat transfer and not in increasing the resistance to heat flow as in insulation.

If heat is passed on from molecule to molecule by heat vibration between them, the process is called conduction. Those materials of high density such as the metals or heavier solid materials have a much higher conductivity than those in which the molecules are far apart and widely separated as with glass fibers, diatamaceous earth or cork.

HEAT INSULATORS

Traditionally, "dead air space" is an excellent insulation against heat transfer. Unfortunately, however, only a few instances can be found in the construction of large rooms and buildings where space insulation is actually effective as "dead air" space.

Dead air space, to be highly effective, must be truly "dead" and prevent any appreciable flow of air. This makes it necessary to fill up the air spaces with very small, lightweight, cellular units if satisfactory heat flow retardation is to be obtained, and the convection losses reduced to a minimum.

Usually the space given to the insulation can be most readily lowered in convection losses by filling the air space with cellular materials of low conductivity, thus shutting off the convection currents and most appreciably decreasing the conductivity factors. Such cellular materials are made up of millions of small air vaults with fibrous walls. These cellular vaults are so small that they prevent air movement and approach most nearly the ideal of a dead air space for insulation.

70

REQUIREMENTS OF GOOD HEAT INSULATION

The characteristics desirable in heat-insulating materials are: (1) low conductivity; (2) good handling characteristics; (3) resistance to decay, deterioration, and odor absorption; (4) extreme lightness to lower the weight in construction design; (5) ability to retain volume dimensions; (6) resistance to moisture; and (7) nonflammable insulation.

Low Conductivity

Usually this is expressed in terms of the number of British thermal units experimental tests have shown will pass through a piece of the material 1 sq ft in area and 1 in. in thickness over a period of 1 hr for each degree Fahrenheit difference in temperature of the 2 faces 1 in. apart.

Such conductivity measurements assume dry materials and as little as 5% of moisture will reduce the effectiveness of the insulation fully 50%.

Good Handling Characteristics

Many of our insulating materials are very hard to handle efficiently. This factor should always be given consideration in selection. For example, cork pipe covering is molded to shape at the factory and is readily applied in a short period of time after it is received by the purchaser. On the other hand, many of the wool felt and rock wool products must be built up on the piping on the job and require considerable time and effort.

Resistance to Decay, Deterioration, and to Odor Absorption

The installation of a good insulation job requires considerable labor and expense. The interest, repair cost, and annual depreciation are of considerable consequence in making the business produce a profit. The insulation selected should be of a permanent nature. If it has any attraction for rodents or vermin, or readily absorbs odors of the produce stored, its use should not be permitted.

Extreme Lightness to Lower the Weight in Construction Design

Fortunately, most of the best heat insulators are very light in weight. Conductivity is closely proportional to the density. The characteristics of lightness and good insulating properties are very closely related.

Ability to Retain Volume Dimensions

But volume dimensions introduce a more serious problem. Such heat insulators as sawdust, shavings, and loosely packed materials settle down year by year leaving uninsulated gaps in the walls.

Molded forms of insulation are usually the most satisfactory, unless ample provision is made to readily repack the walls where loose packing materials are used and subsequently settle down.

Resistance to Moisture

As was indicated under the subject of conductivity, moisture reduces the effectiveness of the insulation very materially. But the resistance to moisture also seriously affects the life of the insulation. When the waterproofing binding of a piece of compressed cork-board begins to deteriorate the board soon disintegrates. Likewise, when rock wool, or similar products, lose their moisture-resisting properties they soon become worthless.

Nonflammable Insulation

The insulation of a wall or pipe with flammable material should be avoided. The insulation should be fireproof to prevent the spreading of a fire that has been ignited near the warehouse walls or pipes.

There are many excellent heat insulators that pass all tests except that of inflammability for most installations. Flammable insulation of either the loose-fill or molded types should be avoided.

PROPERTIES OF HEAT INSULATORS

The most complete piece of work that has been done in determining the properties of heat insulators is that of the US Bureau of Standards. The A.S.R.E. (now ASHRAE) has published a very complete tabulation of the data prepared by different investigators including the Bureau of Standards.

In very cold climates, the alternative might be to put in one positive vapor seal at some midpoint in the wall insulation, then provide some ventilating means to evaporate out any moisture that tends to condense out on either side of this barrier.

The emphasis herein is *positive vapor barrier* wherever placed. Such vapor barriers are limited to the completely sealed heavy permanent plastics, or to continuous metallic sheets. Moist air carrying moisture will penetrate cracks and pinholes since it moves in as a gas, but once the moisture condenses as water inside such a "barrier" the surface tension may prevent the liquid from escaping by the same route that it came into the wall. Water vapor will enter a much smaller pinhole or opening than liquid water can flow out.

This movement of ambient air by the vapor pressure caused by the difference in temperature between the outer hot wall surface and the inner refrigerated insulation through even a small pinhole is not usually understood by installation workmen. When a completely installed vapor seal is provided by the insulation inspector or contractor, any workman who rips, punctures or cuts even a small opening in that vapor seal reduces its eventual effectiveness almost to nil. The air will get in with its load of vapor moisture, then lodge within the insulation as minute drops of water. While a small pinhole will afford more resistance than a torn or ripped opening, it is only a question of time when both will be equally effective in destroying the insulating quality of product.

HOW TO FIGURE INSULATED WALL HEAT TRANSFER BY CONDUCTION

It is customary in North America to determine the transfer rate for each degree Fahrenheit of difference of temperature inside and outside the wall and then find the Btu per square foot per inch of thickness (unit answer) for the composite wall as if it were made of one material. The resistance to the transmission of the heat of one material layer of the wall can be added to the resistance of other wall layers to obtain the total resistance in Btu sq ft per in. of thickness. As in electrical resistances of a single circuit these several resistances can all be considered "in series."

If U is the overall coefficient of heat transfer by conduction and R is the total heat transmission resistance of all of the several resistances in series, then

$$R = \frac{1}{U} \quad \text{or} \quad U = \frac{1}{R}$$

For example, if the wall is made up of a composite section including 8 in. of concrete, 4 in. of cork and 1½ in. of facing cement, then there would be 5 resistances including the outside and inside film resistance at the wall surface, and the resistances of the 8 in. of concrete, the 4 in. of cork board and the 1½ in. of facing cement.

For the concrete, cork board and the facing cement, made up in each case of homogenous materials, the value of the resistance in each case will be the summation of the tabulated value for one inch in thickness of these respective materials.

For computation the abbreviations used are as follows:

> k = thermal conductivity of 1 sq ft taken from the tables for 1 in. of thickness of the homogenous material
> f = the film resistance which for still air (inside) can be taken as 1.65 and for 15-mile per hr wind (outside) can be taken as 6

If k for facing cement is 1.01, for concrete 6.27 and for cork board 0.328 then

$$U = \frac{1}{R} = \frac{1}{\dfrac{1}{1.65} + \dfrac{1.5}{1.01} + \dfrac{8}{6.27} + \dfrac{4}{.328} + \dfrac{1}{6}} = 0.064$$

In this equation the sum of the several resistances making up R include $\frac{1}{1.65}$ for the inside film factor, $\frac{1.5}{1.01}$ for the 1.5 in. of finishing cement, $\frac{8}{6.27}$ for the 8 in. of concrete, $\frac{4}{0.328}$ for the cork board and $\frac{1}{6}$ for the outside film resistance in a 15 mile per hr wind.

The symbol U represents commercially then a unit of resistance for a given type of wall or ceiling. For example, the Federal Housing Administration may specify the walls of a given type of house must have a U value below 0.10 for air conditioned brick homes. The U value for typical cold storage houses is usually specified as below 0.06 and for freezer storage rooms below 0.04.

The air film factors have been determined by experiment and represent the resistance of the air film on the wall surface under consideration. On the outside with higher wind velocities the film thickness is less than on the inside with no air movement and normally has a thicker film of stagnant air.

For emphasis it may be repeated

$$U = \frac{1}{R} \quad \text{the overall resistance}$$

$$C = \frac{1}{R_c} \quad \text{the resistivity to a composite section such as a concrete block}$$
with air pockets molded therein. It is expressed in Btu per hour per square foot per degree Fahrenheit for its total thickness.

$$f = \text{the film resistance} = \frac{1}{R_f}$$

a $=$ the thermal resistance of an air space like C expressed in Btu per hour per square foot per degree Fahrenheit for its total thickness normal to the heat path $= \dfrac{1}{R_a}$

k $=$ thermal conductivity in Btu per inch of thickness per hour, per square foot per degree Fahrenheit $= \dfrac{1}{R_k}$

Thus for any wall, the general conductivity equation is:

$$U = \cfrac{1}{\dfrac{1}{R_f} \text{ (inside)}} = \cfrac{\text{Thickness}}{R_k} + \dfrac{1}{R_c} + \dfrac{1}{R_a} + \dfrac{1}{R_f} \text{ (outside)}$$

COLD AND HEAT TRANSFER THROUGH WALLS AND SPACE

In discussing refrigeration and the insulation of refrigerated rooms or buildings it is of primary interest that a barrier is provided against the heat and moisture traveling by either conduction, convection, or radiation into the insulation to be protected. When the temperature is higher on the outside of an enclosure than on the inside, then the primary interest is forming a thermal barrier against this heat and moisture moving inward.

When the temperature on the outside is colder than the room or enclosure to be protected on the inside, then it is important to establish a barrier against the heat and moisture on the inside moving outwardly through the insulated space and walls.

By one of the basic laws of thermodynamics, heat moves from a condition of high temperature to one of low temperature but never from an area of low temperature to one of high temperature.

In many areas of extreme cold weather it is necessary to protect cold-storage rooms against the cold on the outside to insure the produce against getting too cold within. It is not uncommon to witness northern gardeners engage and use cold storage space in which they store their vegetables to protect them from freezing. In the air cooling of homes it is just as essential to construct the walls and enclosure against the migration of heat and moisture from the inside to the outside as it is to protect the inside against heat and moisture moving inwardly from out of doors.

Since hot air can always carry much more moisture than cold air, it follows that when hot air moves into a cold space the excess moisture

TABLE 6.1

PROPERTIES OF MATERIALS USED FOR HEAT INSULATION

Material	Density Lb per Cu Ft	Average Btu Passing Per Hr Through a Plate of Material 1 Sq Ft in Area, 1 In. Thick, per °F Difference in the Two Faces	Authority
Seaweed, including air	—	0.1788	Rubner (1895)
Kapok, hollow vegetable fibres, loosely packed	0.874	0.238	Bureau of Standards
Cotton, compressed	6.3	0.2061	Randolph
Wool felt, flexible paper stock	20.6	0.363	Bureau of Standards
Cork, ground; grains less than 1/16 in.	9.36	0.2963	Bureau of Standards
Cork, granulated	4.867	0.3225	Biquard
Cork-board, "Non-pareil" (has no artificial binder) . thickness tested 2.03 (in 1917).	9.74	0.305	Willard and Lichty
Cork-board, with bituminous binder	15.6	0.3513	Bureau of Standards
Celotex, insulating lumber 7/16 in. thick	13.5	0.33	George F. Gebhardt
Cabot's car quilt No. 1 eel grass enclosed in burlap 0.83 thick	15.6	0.3193	Bureau of Standards
Peat, ("Torfmull") dry	11.85	0.4074	Nusselt
Lith board (waterproofed): mineral wool; vegetable fibres & waterproofing binder. Rigid.	12.5	0.3802	Bureau of Standards
WOODS			
Planer shavings	8.74	0.4063	Bureau of Standards
Sawdust, various	11.86	0.4063	Bureau of Standards
Cypress, across grain	28.7	0.668	Bureau of Standards
Pine, across grain	—	0.2555	Forbes
Yellow pine, across grain	—	1.045	Bacon
Oak, across grain	38.0	1.016	Bureau of Standards
White oak, across grain, thickness of sample 0.516 in.	37.45	1.3215	Taylor
White oak, along grain, thickness of sample 0.754 in.	37.45	2.74	Taylor
Maple, across grain	44.3	1.1025	Bureau of Standards
ARTIFICIAL WOODS			
Cement wood (sawdust and Portland cement) dry	44.6	0.9283	Nusselt
MINERAL MATTER			
"Calorox" fluffy, finely divided mineral matter	4.0	0.2207	Bureau of Standards
Lampblack, Cabot's No. 5	12.05	0.2679	Randolph (1912)
Mineral (slag) wool, loosely packed	12.0	0.2614	Bureau of Standards
medium packed	12.5	0.2760	Bureau of Standards
"Rock Cork" mineral wool binder and waterproofing; rigid; high density	15.6	0.328	Bureau of Standards

TABLE 6.1—(*Continued*)

Material	Density Lb per Cu Ft	Average Btu Passing Per Hr Through a Plate of Material 1 Sq Ft in Area, 1 In. Thick, per °F Difference in the Two Faces	Authority
ASBESTOS			
Asbestos paper: built up of thin layers, with organic binder	31.2	0.494	Bureau of Standards
Asbestos paper	—	1.249	Lees & Chorlton
Asbestos fire felt, very light but self-sustaining	7.24	0.4437	Randolph
Magnesia, 85%, and asbestos (no forced drying)	13.48	0.5104	Randolph
J.M. asbestocel pipe covering layers of plain and corrugated (1/8 in. deep circular) asbestos paper around pipe for medium pressures.	12.06	0.596	McMillan
Asbestos, hand-packed, hard	43.08	1.525	Groeber (1910)
Fire-felt roll	26.2	0.596	Bureau of Standards
Sil-O-Cel, infusorial earth; natural blocks	28.1	0.5805	Bureau of Standards
Fuller's earth, argillaceous powder	33.0	0.697	Bureau of Standards
Blast furnace slag, dry grains 0.08 in. to 0.20 in.	22.48	0.718	Hencky
Quartz, sandfine	—	0.3805	Forbes
Quartz, coarse	96.75	2.084	Randolph
Boiler clinkers, dry	46.8	1.089	Hencky
Gypsum plaster board, such as "Sheet Rock" & "Adamant" covered with paper about 0.02 0.02 in. thick Thickness 1/4 in. to 1/2 in. called fire-resistive wallboard 32 × 36, or 48 in.	61.0	2.6	Herter (estimated)
Sand, fine grains, smaller than 0.08 in. dry	99.8	2.146	Desvignes
Sand, river, fine-grained, dried completely by heat	94.8	2.340	Groeber
Facing cement, as used for plastering cork slabs (magnesium oxychloride composition)	—	1.010	Griffiths
Portland cement	—	2.061	Lees and Chorlton
Plaster, ordinary, mixed, dry	83.55	2.422	Desvignes
Soil, dry	—	0.958	Lees and Chorlton
Cement, mortar, Portland	117.6	3.71	Desvignes
Concrete, stone mixture 1/2/5	—	6.277	Chas. L. Norton
Asphalt roofing	55.0	0.697	
Roofing covered with gravel	—	1.325	
Glass, flint	—	4.16	H. Meyer (1888)

TABLE 6.1—(*Continued*)

Material	Density Lb per Cu Ft	Average Btu Passing Per Hr Through a Plate of Material 1 Sq Ft in Area, 1 In. Thick, per °F Difference in the Two Faces	Authority
MASONRY			
Bricks, very porous, dry	44.3	1.17	Hencky
Stone, soft	91.75	3.47	Biquard
Bricks, American machine made in one course, wall 3.79 in. thick	131.9	4.14	Willard and Lichty
Hollow tile masonry, dried 6 months	—	2.29	Groeber (cube)
Haricloth	—	0.1167	Forbes (1872)
Horsehair, pressed	10.74	0.12	Nusselt
Feathers, with air	—	0.1667	Rubner
Felt	—	0.2525	Forbes
Sheep's wool, slightly greasy	5.24	0.5348	Desvignes
Wool felt	9.36	0.508	Taylor
Steel wool, No. 2	6.86	0.596	Randolph
Steel, carbon 1%	—	322.4	Lees, Lees, Jaeger & Diesselhorst
Copper	—	2094.0	Lorenz

which the cold air cannot hold will condense out into or on the walls or pipes. Such moisture condensing within the walls makes the insulation nearly worthless as a heat barrier.

THE PROBLEM OF ADEQUATE INSULATION AND PERMANENT VAPOR BARRIERS

It is essential to a satisfactory solution of providing adequate permanent insulation and equally permanent vapor barriers in walls and ceilings of enclosures and rooms that the designer and installer should be fully aware of the inherent difficulties and seriousness of the job. Such designing requires real ingenuity on the part of the responsible engineer of the project.

One of the greatest contributions members of the refrigerating and air conditioning engineers societies can make to their associated industries is to improve the art of producing and installing a practical permanent vapor barrier for room insulation that will keep the moisture from entering the insulation from the out-of-doors in the summer and also prevent the moisture from getting into the insulation from the inside rooms in winter.

Most vapor barriers for cold storage rooms are installed on the outer wall side to protect the installation from moisture moving into the colder insulation, then condensing. If the outside temperature never drops lower than the inside, then virtually no moisture will travel from the cold room to the insulation and no vapor barrier is provided on the cold side of the wall. Fortunately, the amount of moisture per cubic foot of cold storage room air is very small and even if the outside temperature drops lower than that of the cold room, the condensation in the insulation is not too serious.

But if instead of a cold storage room, the enclosure insulated is a warm kitchen where steam from cooking is plentiful, when winter comes there will be a heavy movement of moisture moving outward into the insulation if no inner wall moisture barrier is provided. Thus in an enclosure that must be heated in winter and cooled in summer for either air conditioning or storage of perishables, the real need is for a positive vapor seal on the inner walls for winter and another positive seal on the outer wall for summer.

RECAPITULATION ON VAPOR BARRIERS

The major significant detail of a refrigerated installation is the proper selection and installation of the vapor barrier.

All insulators should be installed with vapor barriers.

The vapor barrier should be located on the *exterior* surface of all insulation cavities and on the warm side of partitions having significant temperature differentials. It is not to be construed from this that a vapor barrier would be required on the interior surface of refrigerated spaces in arctic regions where the outside temperature may be lower than holding temperatures in the refrigerated space.

The preferred vapor barrier material is metallic foil, laminated between layers of protective coverings. Observations in this study have been limited to aluminum metallic foils.

Mop coats of bitumens should not be considered vapor barriers. However, these and similar coatings within the vapor barriers envelope may form vapor checks to the extent that condensation occurs if located in the dew point range or ice may form if located in freeze point range. These vapor checks should be avoided.

All joints or seams of the vapor barrier should be lapped and sealed. The preferred means of sealing is the use of pressure-sensitive tape, similar in characteristics to the vapor barrier material; staples or nails should not be used, but when the use of nails or staples to attach vapor barriers cannot be avoided then the penetrations should be sealed with pressure-sensitive tape.

Acceptable vapor barriers are those which satisfactorily pass the flat-sheet and creased-sheet tests. The sample for the creased sheet test should be prepared in accordance with ASTM Standard D-1027-51 by either method A or B. The measurement of water vapor transmission of material in sheet form should be conducted on the basis of ASTM Standard E-96-53T, Procedure E. The flat sheet should exhibit no greater than 0.01 perm value after test, and the creased sheet should exhibit no greater than 0.02 perm after test.

The installed vapor barrier, with seals, should be of such design and nature that it can absorb expansion and contraction in the order of magnitude of 1 in. per 100 ft, remain in a resilient state at operating temperature, and be resistant to chemical attack. The sealants used should be compatible with the insulation and vapor barrier.

The vapor barrier should be completely installed prior to application of the insulation. The vapor barrier should then be carefully examined by the owner's representative to insure that the material is properly installed and sealed. At this time, all necessary openings in the vapor barrier should be prelocated and provision made for their proper sealing.

Vapor Barrier Comments

A good vapor barrier, properly installed, will provide many years of economical low-temperature insulated storage with most commercial low temperature insulants. The hazard of deterioration due to moisture is greater with organic insulations and organic structural members because of the difficulty of avoiding vapor leaks in a vapor barrier, as well as the nature of organic materials.

The National Academy of Sciences survey teams found only one type of insulation; cellular glass free of moisture except on its surface. All other insulating materials were found to contain some moisture.

The average moisture content of the organic-board-insulation core samples is approximately 30% by weight, while the average moisture content of glass-fiber core samples was approximately 18% by weight and no moisture was found in the cellular glass insulation.

If a positive vapor seal is installed, most any of the commercial cold storage insulants can be used with success, but if a defective vapor seal exists as a protection for most available insulations, it would have been better to not install any of the insulants except probably glass fiber.

INSTALLATION FAILURES OF INSULATION AS REPORTED BY THE NATIONAL ACADEMY OF SCIENCES SURVEY TEAM

Installation failures and weaknesses with time reveal little appreciation of the effect of mechanical penetration of the vapor barriers. In the hands of uninformed builders, equipment hangers, meat racks, sprinkler systems and electrical wiring have been found to have pierced well designed barriers at random even at time of construction. In some cases these metal penetrations were not insulated and themselves became notable heat carriers. Failures of an entire wall or ceiling often results from such practice.

The installed continuous defrost installation systems vary in type from those designed for manual removal of frost. Manual frost removal has been found most unsatisfactory since it often takes up to two weeks to complete single rooms. Preferable installations include hot gas or electric continuous defrost systems, automatically controlled, and are widely used with good results. For large freezer spaces using multiple evaporators, sectional defrost is widely favored to avoid large temperature increases within the space. Freezing of defrost drain lines is a common complaint; electric heater cables are generally used to correct this condition, with satisfactory results.

Frequent and extensive cracking of exterior walls was found in many installations. It is quite obvious from the survey results that these problems are significant to the designer and management.

Refrigerator doors were found to be a universal source of complaint. Such problems as freezing, sagging, rotting of wood members, and external physical abuse have been repeatedly noted. Few managers are satisfied with air-operated doors; almost all prefer doors that are electrically controlled. Some comments were received to the effect that, with the trend of power-driven trucks and larger pallets, traditional doorways are proving insufficient in width and height. Vestibule doors are generally used for freezer spaces where traffic is frequent. "As-installed" drawings were noted to be inaccurate and often specifications not followed by the builder. Generally, design drawings were inadequate in describing the various components used and the techniques of installation.

NIGHT COOLING EFFECTS ON COLD STORAGE INSULATION

Evidently some of the unexplained top floor insulation failures are brought about by night cooling (sometimes called interstellar cooling) of the roof at which time the vapor movement may be reversed in cold storage rooms from inside by a lower temperature on the outside roof surface than within the top floor rooms. Under recurring low outside roof surface temperature conditions, the upper room moisture with pressures reversed will penetrate the insulation from the room ceiling within, unopposed by any vapor barrier. Such moisture lodges in the insulation and destroys much of the future resistance to the incoming heat flow. Cold storage rooms beneath an insulated vapor barrier constructed roof are more susceptible to this reversal of vapor flow than freezer storage space since outside temperatures plus the increment of

TABLE 6.2

INSULATING SLABS OR BOARDS

Material	Density Lb per Cu Ft	Average Btu Passing per Hr Through a Plate of Material 1 Sq Ft in Area, 1 In. Thick, per °F Difference in the Two Faces
Cellular glass	9	0.41
Glass fiber	7	0.21
Polyurethane (exp.)	3	0.17
Rubber (exp.)	4.5	0.22
Polystyrene (extruded)	1.9	0.22
Expounded polystyrene	1.0	0.24

interstellar cooling seldom would total lower than the existing freezer storage condition while a cold storage temperature may be much higher within than without.

The air space under the roof should be kept well ventilated by adequate air movement to prevent condensation occurring from interstellar or night cooling within this space. This condensed moisture may act as "rainfall" upon the insulated ceilings beneath with the telltale moisture stains appearing in the top floor storage or processing rooms.

ECONOMICS OF INSULATION INSTALLATIONS

In the installation of insulation, the builder of a cold or freezer storage house should determine the lowest long-term ownership cost per wall, floor or ceiling by his thermal design. The economic maintenance of refrigerated room temperature over a given period of years of usefulness should be balanced against the capital cost of the insulation of the building shell, and the annual cost of operation including the power to maintain the necessary refrigeration for the hypothetical different thicknesses of insulation. The total cost of each wall, floor or ceiling should include the cost of producing the necessary cold to compensate for both the heat gain and for the fixed charges of installation.

Emphasis should be placed on the annual charges for both the insulation, including interest, maintenance and depreciation, and for the power cost of producing each ton-year of the required refrigeration.

In making these computations the maximum return on the investment will apply to the overall cost of maintaining the temperature of the storage room including insulation of the ducts and of all other costs of the cooling load involved in the installation.

TABLE 6.3

COLD AND FREEZER WAREHOUSES MINIMUM INSULATION THICKNESSES

Storage Temperature °F	Insulation k = 0.3 Thickness In.	
	Northern US	Southern US
25 to 40	4	5
15 to 25	5	6
0 to 15	6	7
0 to −15	7	8
−15 to −40	9	10

STRUCTURE FAILURE OBSERVATIONS WITH INSULATION AS REVEALED BY A NATIONAL CHECK SURVEY OF COLD AND FREEZER STORAGE WAREHOUSES

In almost every case the moisture content of ceiling samples is significantly greater than that of wall samples.

The field survey shows that a few installations seem to have performed quite satisfactorily for many years, but a large majority have definite and serious deficiencies.

No termite infestation was detected in any of the installations checked, although in two cases termites were given as the reason for rehabilitation. Cases have been observed by others where termites were lifted by the warehouse elevator from an infested storage yard or shed. The needed moisture, cellulose food and desired darkness were furnished the termite within the existing structure where they were established.

A few cases of rodent nesting were found in the organic insulations, but examinations revealed no eating of the insulation.

In the majority of cases, galvanized metals, especially when used in light fixtures, were corroding.

TABLE 6.4

HEAT GAIN FACTORS (WALLS, FLOOR AND CEILINGS) BTU PER SQ FT (24 HR)
COLD STORAGE WAREHOUSES

Insulation k = 0.3 Thickness In.	Temp Difference (Ambient Temp Minus Storage Temp), °F								
	1	40	50	60	70	80	90	100	110
4	1.8	72	90	108	126	144	162	180	216
6	1.2	48	60	72	84	96	108	120	132
8	0.90	36	45	54	63	72	81	90	99
10	0.72	29	36	43	50	58	65	72	79
12	0.60	24	30	36	42	48	54	60	66

The exteriors of refrigerated structures have proved surprisingly porous. Operational difficulties are more evident on the windward side of the building. In several cases, the addition of breakdown areas and similar areas on the windward side has materially reduced refrigeration difficulties.

In only a few instances were wood beams found in good condition. Wood has been found warped, buckled, shrunken, bulged and in a few cases termite-infested. An exception is wood used in conjunction with reflective insulation in freezer applications which appear to be in good condition. Heart redwood, tidewater cypress, and pressure-treated woods appear to be least liable to deterioration.

A number of serious floor heavage problems have been observed under freezers. The most satisfactory method of correction has been the removal of the floor and installation of heating pipe coils beneath the floor.

GENERAL SUGGESTIONS ON INSULATION

Nonhygroscopic insulations of inorganic or plastic materials, board or batt type, should be used for both freezer and cooler space.

Where natural organic, hygroscopic insulations are used, "k" factors should be based on average moisture content, and the thickness of insulation increased to hold the design "U" value of walls, ceilings, and floor. A minimum moisture content of 15% by weight (approximately 1.6% by volume) should be used in these calculations. Loose-fill insulations are acceptable.

Reflective insulation should have a minimum reflectivity of 95% on both sides of the sheet.

The owner's inspectors should be experienced and trained in the field of refrigerated structures. Plumbing, electrical, and general construction inspectors are not acceptable substitutes.

Insulations that are stored on the job site should be protected from the elements and physical damage and termite invasion.

The use of wood within the vapor-barrier envelope should be held to an absolute minimum.

The roof problem should be segregated from the ceiling problem. The preferred design procedure is to provide a ventilated air space between the refrigeration insulation and the roof; or, where a concrete slab is used, the vapor barriers and insulation should be located on the interior surface of the slab.

Ceiling areas in which air space is provided above the vapor barrier should be ventilated. Where large latent heat loads may be produced (as in kitchens), provision should be made for mechanical fresh-air ventilation.

OPERATIONAL PRACTICES

Wattmeters should be installed on each cold-storage installation for use as trouble indicators. Comparison of power consumption, as installed and in operation, can be used to check plant efficiency and locate possible difficulties.

Cold-storage spaces should be brought down to operating temperatures in slow stages during initial start-up. Rapid decreases in internal pressures may cause serious damage.

BIBLIOGRAPHY

ANON. 1966. Cold store on floating raft. Mod. Refrig. 69, No. 823, 873, 875–876.

ARMSTRONG CORK Co., 1900. Low temperature insulation core sampling procedure. Lancaster, Penn.

ASHRAE. 1959. Heating, Ventilation and Air Conditioning Guide, 37th Edition. Am. Soc. Heating, Refrig. Air. Cond. Engr., New York.

ASRE. 1956–1957. Data Book, 6th Edition. Am. Soc. Refrig. Eng., New York.

BUILDING RESEARCH INSTITUTE. 1959. Sealants for curtain walls. Natl. Acad. Sci., Washington, D.C.

DEPARTMENT OF THE AIR FORCE. Cold storage warehouses manual AFM-88, 21.

DILLON, R. M. 1960. Refrigerated storage installations, Publ. 759. Tech. Rept. 38. Federal Construction Council. Natl. Acad. Economics.

FOREST PRODUCTS LABORATORY. 1955. Wood Handbook 72. U.S. Dept. Agr.

HALL, H. E., FORD, P. J., and THOMPSON, K. 1966. Helium 3 dilution Refrigerator. Cryogenics 6, No. 2, 8008.

HOLMGREN, J., and ISAKSEN, T. 1959. Ventilated and unventilated flat com-port roofs, Norwegian Building Res. Inst., Oslo, Norway.

HOOPER, W. F. 1959. Method of measuring the odor absorption and retention properties of surfaces. ASHRAE Trans. *65*, 735–744.

KAYAN, C. F., and GATES, R. G. 1958. Influence of insulation on moisture-condensation aspects of steel-framed cold storage warehouse structure. Refrig. Eng. *66*, No. 1, 39–44.

KROPSCHOT, R. H. 1959. Cryogenic insulation. ASHRAE J. *1*, No. 9, 48–54.

LaJOY, M. H., and WHITNAH, G. R. 1953. Heating Arctic Buildings. Heating, Piping and Air Conditioning, Chicago, Ill.

MORPHEW, A. R., JONES, A. J. 1967. Safety in Cold Stores. Mod. Refrig. *70*, No. 829, 85–86, 397.

PENNINGTON, C. W. 1967. How insulating glass cuts cooling loads. Florida Eng. Ind. Expt. Sta., Gainesville, Fla.

SIMONS, E. 1955. In-place test studies of insulated structures. Am. Soc. Refrig. Eng. *63*, No. 2, 40–47, 100–102, 106, 110, 112, 114.

WHIPPO, H. M., and ARNBERG, B. T. 1955. Survey and analysis of the vapor transmission properties of building materials, *PB131219*. US Dept. Comm., Office Tech. Serv., Washington, D.C.

WOODSIDE, W. S. 1948. Cold room insulations. Am. Soc. Refrig. Eng. *56*, No. 3, 225–226, 268.

WOOLRICH, W. R., and MEI, H. T. 1959. Specifying vapor barriers. Air Conditioning, Heating and Ventilation *59*, No. 9, 55–58.

CHAPTER 7

The Insulation of Freezer Warehouse Floors

THE ECONOMICS OF INSULATING FREEZER FLOORS

A question that arises in almost any discussion pertaining to freezer storages is the feasibility of placing freezers on grade, or fill, with no commercial insulation used in the floor as a heat barrier. Such a floor uses the insulating effect of the soil beneath it as an insulant rather than an added or commercial type insulation. A rather good case can be made either for or against this type of floor construction and it is very probable that no complete answer exists. This chapter is an effort to present various arguments both for and against the so called noninsulated floor as well as arguments for or against an insulated floor. Each floor location can cause different problems that must be solved. Unfortunately, the complexity of the problems and divergent views preclude a simple answer and the final analysis is usually an educated guess using past experience and available knowledge to determine the best course of action.

NONINSULATED FLOORS

Definition

It is well to define a so called noninsulated floor that will be discussed in this chapter. This floor can be defined as a freezer floor, usually concrete, resting directly on grade, or on a fill, with no commercial grade insulation installed as a barrier to the flow of heat, but utilizing rather the insulating effect of the soil, or fill, as a heat barrier. In no instance should the term "noninsulated" be taken to mean a floor of concrete or other structural material over a void. With this in mind, the discussion of noninsulated floors becomes actually a discussion of native earth or fill as an insulant. Since earth substances come in infinite varieties, it can be seen that there can be infinite solutions to the problem which, in reality, means no solution other than the educated guess.

Sand, clay and other types of ground, all have certain insulating values. In some common soils, between 2 and 3 ft thickness of soil will be equivalent to 1 in. of good commercial insulation. Thus, within limits, there is good insulating value in a "noninsulated" floor if the ground below the floor is taken into consideration. The insulating value of the soil will vary rather widely depending on the soil structure, proximity above free water and other factors. In a good dry area, it will generally be found that, after stabilization of a floor, the heat loss through the floor will be about the same as if there were 6 in. of good commercial insulation used. This will be an average figure as the actual losses will vary in various parts of the room. At the edges of the room around the perimeter, there will be high losses to adjacent warm areas unless a band of insulation is extended into the room or trenched down from the wall. Noninsulated floors seem to lend themselves better to large areas than small ones. If small rooms are to be used without floor insulation, it is almost mandatory that a band of insulation be used around the perimeter of the room, either extending in 2 or 3 ft or down below the walls to break the flow of heat in from warm adjacent spaces. In considering floor insulation purely from the standpoint of heat losses, it is doubtful that the cost of commercial insulation could ever be justified.

Heating of Noninsulated Floors

Experience has dictated that some means of heating should be provided beneath a noninsulated floor. If a room is used as a freezer for an extended period of time, the temperature of the soil beneath the floor will fall gradually and the freezing line will deepen with the passage of time. Observations taken of a freezer room held at about 0°F over a period of several years indicated that the freezing line under the floor was at a depth exceeding 35 ft and may have extended as far down as 50 ft. The advance of the freezing line is progressive and is apparently at a slowing rate as it deepens but advances to considerable depths when no heat is available. As the freezing depth beneath the freezer floor increases, ice lenses can and do form but not always at the same rate or thickness. Much research could be done on the formation of these ice lenses to determine just what type of soil and other conditions increase or retard their formation. As an ice lens forms, it tends to block off the migration of moisture from the ground to the room cooling coil by stopping the moisture at the lens. This, in turn, causes a thicker lens to form and as a logical extreme, the ice expansion can cause floor heavage. There is no known rule or rules to predict floor heavage. In moisture laden areas, it can occur fairly soon while in relatively dry areas it may not occur for a number of years and with dry soil and good porosity in the floor slab, heavage may never occur. The problem that occurs is that the condition of the soil under the freezer floor can change; sometimes with the seasons of the year and at other times as a result of changes made by building or excavation even at some distances from the actual freezer room. This is the reason that hard and fast rules are difficult if not impossible to formulate. The ideal noninsulated floor structure would consist of a very porous wearing floor and soil conditions under the wearing floor that would allow free passage of water vapor to the cold room coils where it would condense on the coils in the form of frost. If the substructure will allow passage of vapor as fast as it forms at the deeper levels and it can pass through the wearing floor in the form of vapor to the cooling coils; then heavage probably will not occur.

Stopping Floor Heavage

Floor heavage can usually be stopped but the process of heaving will not, as a rule, reverse itself. The best solution to floor heaving is to prevent it from starting. This can be done by introducing a heat source at some point below the surface of the floor. The heat source can be almost any type of relatively mild temperature heat. Sometimes pipe coils are placed below the surface of the floor when the room is constructed. Oil or other nonfreezing liquid is pumped through a heater and through the coils to carry heat to the underfloor area to prevent the formation of ice. A good heat source for this type of heating is the discharge gas from the refrigeration compressors. The discharge gas is routed through a heat exchanger before entering the condenser and the heat necessary is extracted and transferred to the solution in the heat exchanger which is then circulated through the coils beneath the freezer floor. Electric heaters are also satisfactory and are also relatively easy to apply to an existing floor. Actual depth required for heaters is not certainly known. If heaters are placed at a depth of 3 or 4 ft, it is almost a certain guarantee that harmful ice will not form with normal earth moisture present; but at this depth, a considerable amount of heat will be transmitted from the heaters to the freezer room thus increasing the refrigeration load. Experience indicates that a somewhat greater depth will give satisfactory results with far less heat passage to the freezer room. Electric heaters of the slug-type have been successfully used for this purpose. In a large area, they may be placed at a depth of about 10 ft and on about 10-ft centers. A heater with a power rating of 400 to 500 watts is usually sufficient. It is good practice to use a lower voltage than is rated on the heaters. This allows a lowered surface temperature of the heater, which is desirable, and also greatly lengthens the life of the heater. Three 115-v heaters in series on a 230 v power source will usually give adequate performance and long life. Heater slugs can be obtained with long electrical leads suitable for direct burial. Holes can be drilled for the heater slugs and connecting wiring run in airport type cable in slots sawed in the freezer room floor. Thermocouples placed at various depths and in several locations under the floor are desirable to determine underfloor temperatures and can effect power savings by allowing the heaters to be cut off when not needed. Programs can also be set up to utilize the heaters during off electrical peak periods as much as possible. Temperatures of about 36°F at a 10 ft depth in well-drained areas seem to eliminate any serious threats of floor heavage. With pumped solution in pipe coils, a solution temperature of about 40°F should be satisfactory. Coils should also be placed rather deep to prevent excess heat from entering the room. It will usually be found that after stable conditions are reached, intermittent operation of the heaters will produce satisfactory results. The use of the thermocouples will aid in setting a schedule of heater use.

To Heave or Not to Heave

Observations indicate that, at the present time, there is no certain way to determine whether floors will heave or at what period of time they are likely to heave. If floors are to be used without commercial insulation, heaters installed as the plant is built are certainly a wise investment. They may seldom or never be required in one room; and in another room only a short distance away, may be in almost constant use.

Large Freezer Room Floors

In large cold storage rooms, a noninsulated floor represents a considerable savings in first cost. It is, however, a calculated risk since there are certain inherent hazards in this type of construction. Each case is different and all of the conditions pertaining to an area should be studied prior to the construction of the floor. Obviously such a floor should not be used in coastal areas, or in any area, where free water may be found at shallow depths. This can only lead to ice problems. In a relatively dry area, however, with no underground free water at shallow depths, a noninsulated floor may work out very well. Heat losses are not excessive and cause no particular strain on the refrigeration machinery. Heaters beneath the floor surface at reasonable depths are much cheaper and simpler to install when the room is built rather than after it is in use as a freezer. It is very difficult to drill into a frozen floor and can consume a great amount of time.

Floor Heaving Control

The ideal floor surface where no commercial floor insulation is used would be no surface material at all except the soil. As this is not usually possible or desirable, a concrete floor is most often poured on the ground as a wearing surface. Care should be exercised that no hardeners or other additives be used in the concrete that will tend to vaporproof the floor and form a barrier to the free passage of vapor through the floor to the cold coil surface. If the water vapor can travel freely through the floor from the soil beneath, it will be found that very dry earth will be immediately below the slab with a gradual increase in moisture content as greater depths are reached. If a balance can be reached where vapor can be drawn to the room coils as fast as it enters the soil structure below the floor and there are no vapor barriers in the soil structure, it is probable that heavage can be controlled.

Earth Fly Wheel Effect

It is obvious that it will take much longer to pull a freezer room down to temperature with a noninsulated floor than one with heavy commercial floor insulation. As the room temperature reduces, a considerable tonnage of earth under the floor must also be pulled down in temperature and eventually frozen. For this reason, rooms that are newly pulled down in temperature will gain temperature rapidly when refrigeration is lost. A large room may take 90 days or more to stabilize to be a good working room. The opposite side of the coin is, of course, that after the room has been pulled down and the floor stabilized as to temperature, it is difficult to alter the temperature in a short span of time. The frozen earth mass beneath the freezer room exerts a tremendous flywheel effect on the room so that loss of refrigeration will have very little effect after an initial bounce of 1° or 2°. Even in a 48 hr period with no refrigeration, temperatures will not usually rise over 5° or 6°F.

In calculating losses from a large freezer room without floor insulation, the calculations will not be different than for a room with commercial floor insulation. The main difference in the rooms is the pull down period. The room with the noninsulated floor will require a long period of time to completely stabilize. After stabilization, there will be the flywheel effect which will prevent any sudden change in temperature. Actual observation of a room of some 20,000 sq ft with a noninsulated floor and held at 0°F for several years indicated a very slow warm-up after the refrigeration was turned off. After about a

30-day period with no refrigeration on the room at all, the temperature within the room was still below freezing. Outside temperature averaged about 70°F. The roof was exposed to the sun and there were cooler rooms, above freezing, on either side.

Heavage is always a danger in any room held below freezing when it is on grade and whether it has an insulated floor or not. It may not evidence itself for a long period of time and several years may pass before it shows up and, in some instances, the floor may never heave. No positive test or method is known that can give heaving information before a plant is built and for this reason, it is a wise precaution to install underfloor heating in all floors of freezers that are constructed on grade or on fill.

INSULATED FLOORS

When to Insulate Freezer Room Floors

In many instances there is a definite advantage in the use of insulated floors. This is particularly true in low lying areas and wherever underground water levels are high. Convertible rooms which are converted back and forth at rather frequent intervals from freezer to cooler and vice versa will find the flywheel effect to be detrimental when warming from a freezer to a cooler. Smaller rooms also seem to operate better with floor insulation. It should be emphasized that heavage of insulated floors on grade can occur much more rapidly than noninsulated floors unless proper precautions are taken to prevent the heavage. Insulation, no matter what the thickness, allows some transfer of heat. Thus with the passage of time, an insulated floor on grade will allow the temperature below the floor to drop below the freezing point. This will normally occur if no means of bringing heat to the subsurface is used. Normal earth will form enough insulating effect to prevent the rapid transfer of heat to the underside of the floor and the temperature will drop. As the freezing point is reached, moisture can not be evaporated to the cooling coils because floor sealing is usually used on insulated floors to keep the insulation dry and effective. Since the moisture is trapped beneath the floor, it will freeze into solid ice and expand and cause heavage.

Ventilation Tiles Below Insulated Floors

The heavage beneath insulated floors can be rather easily prevented by supplying a low level heat to the underside of the floor. This can be electric heat or warm fluid in pipe coils or by means of warm air ventilation tiles beneath the floors. The ventilation is the least expensive when it can be done since it requires no auxiliary heating in temperate climates and only a slight amount in cold climates. Only enough heat need be transmitted to hold the underside of the insulated slab to a temperature of 5° or so above the freezing point. Quite satisfactory results can normally be obtained with ventilation tile by using about a 6-in. concrete tile on approximately 6-ft centers across the freezer room and placing the tiles a foot or so below the insulated floor slab. The tile should be pitched for drainage and also as an aid in air circulation through the tile. A freezer room at dock height is an ideal application for vent tile as the tile can be placed in the fill above grade so that it is open at both ends. With good insulation in the floor, temperatures of about 40°F will normally be maintained under the slab by the natural air currents through the vent tiles. It is still possible to use tile even when the freezer floor is not at dock height, but on

Courtesy of Alford Refrigerated Warehouses, Corpus Christi

FIG. 7.1. INSTALLING VENTILATION TILES BENEATH FREEZER FLOOR

Tiles run the full length of the freezer room and are pitched for drainage and air flow purposes. This type installation used with insulated floors at dock height.

grade, by using fans to force air through the tiles. Heating the air before it enters the vents is not usually necessary except possibly in climates where there are prolonged periods of subfreezing weather. Even then, if the vents are closed during cold spells, the heat sink effect of the ground will sometimes act as a flywheel to prevent the subsoil from freezing. Even a short period of the subsoil temperature below freezing is not harmful if circulation is resumed when the air temperature is above freezing.

Thermocouples in Freezer Room Floors

Any floors, insulated or noninsulated should have thermocouples buried at various depths in several locations throughout the room so that periodic readings may be taken to determine the temperatures at various depths. In this way, the timing may be set up for the use of artificial heat so that only the minimum amounts are used.

Courtesy of Alford Refrigerated Warehouses, Corpus Christi

FIG. 7.2. CONSTRUCTION VIEW SHOWING VENTILATION OPENINGS IN DOCK WALL FOR UNDERFLOOR VENTILATION OF INSULATED FREEZER FLOOR

FLOOR DESIGN STUDY

No hard and fast rules can be made as to when it is economically feasible to use an insulated floor or a noninsulated one. Probably rooms of less than 5,000 sq ft area may be better with insulated floors regardless of the soil conditions. Savings in first cost are not too great and in the smaller areas the edge losses of refrigeration are of some magnitude since they are a greater proportion of loss in the rooms of smaller area. This can require considerable amount of extra refrigeration. With insulation either in a band around the perimeter or in a trench under the walls, edge losses are reduced but the initial cost approaches the cost of full insulation.

In larger rooms, considerable study should be given to the use of insulated vs noninsulated floors. Dryness of the area and the structure of the soil are all points that should be taken into consideration. In dry locations, the larger the room, probably the more favorable the use of the noninsulated type of floor construction. Regardless of any soil tests and speculation, it is probably advisable to use heaters of some type beneath the floor. This is relatively inexpensive when done with the building construction and is certainly excellent insurance against future troubles. Insulated floors should also have a means of heat when installed on grade or on fill.

There is really no easy answer to the floor problem for freezers. There are pitfalls and problems regardless of the route followed. Each room is a separate problem and should be studied at some length before a decision is made on what is the best construction for that room.

BIBLIOGRAPHY

ANON. 1947. Remedying frost-heave effects. Mod. Refrig. 50, 133.
ANON. 1948. Subfloor ventilation prevents frost heavage. Refrig. Serv. Engr. 16, No. 3, 10, 20–23.
ANON. 1966. Cold store on floating raft. Mod. Refrig. 69, No. 832, 873, 875–876.
COOLING, L. F., and WARD, W. H. 1944. Damage to cold stores due to frost-heaving. Inst. Refrig. Proc., 41, 37–48.
COOLING, L. F., and WARD, W. H. 1950. Damage to cold stores due to frost-heaving. Ice and Refrig. 119, 1, 41–44.
DART, D. M., CATES, R. E., MILLE, R. B. Heating, Piping and Air Conditioning 38, No. 6, 127–131.
DILLON, R. M. et al. 1960. Refrigerated storage installations. Publ. 759, Tech. Rept., No. 38. Federal Construction Council, Natl. Acad. Economics.

HALLOWELL, E. R. 1949A. The Alford refrigerated warehouses. Refrig. Eng. 57, 226–237, 272–275.

HALLOWELL, E. R. 1949B. Recent developments in refrigerated warehouses. Ice and Refrig. 116, No. 3, 53–56.

LEIDING, O. 1948. Huge Refrigerated Warehouses Bring Model Storage at Dallas. Food Inds. 20, 1284–1288.

STENCEL, R. A. 1947. The effect of frozen foods on cold storage warehouses. Can. Refrig. J. 13, No. 10, 15.

WOOLRICH, W. R. 1966. Handbook of Refrigerating Engineering, 4th Edition, Vol. 2. AVI Publishing Co., Westport, Conn.

CHAPTER 8

Machine and System Selection for Small and Intermediate-Sized Storages

HISTORICAL

Not many years ago, there was very little choice in the selection of a refrigeration system for the small and intermediate-sized plant. The existing systems used ammonia as a refrigerant, had water cooled condensers and the evaporator consisted of pipe coils with direct expansion ammonia or used brine in the coils chilled by the ammonia plant. Ammonia machines were made in very small sizes to take care of small loads as encountered in butcher shops, etc. Control was more or less rudimentary and not too reliable. Small areas such as grocery cases and drug store boxes of the period used ice as a refrigerant. Ice cream cabinets were iced down with a mixture of ice and salt. The self-contained refrigerated cabinets, as we know them today, did not exist.

With the advent of methyl chloride and sulfur dioxide as refrigerants, the use of ammonia for the small refrigeration plant pretty well faded out of the picture. Both of the new refrigerants permitted the use of air cooled condensers, thus eliminating the cooling tower or direct use of city water. First cost savings with the new gases were considerable since the equipment was much lighter weight than used with ammonia. Copper lines could be used which speeded up erection procedures in the installation of the small plants. The expanded use of the small condensing unit capabilities coupled with changing pattern of food distribution and the advent of frozen foods combined to cause a phenomenal growth in the industry in the space of only a few years.

Sulfur dioxide and methyl chloride were both desirable from a refrigerant standpoint and performed well but both of them presented hazards from the safety standpoint. These two refrigerants had not been in use long before the advent of the halocarbon refrigerants. There is no question that the halocarbon refrigerants caused a tremendous spurt in the building of refrigeration systems. Refrigerant 12 was the first of these refrigerants to come into wide use with the other ones following within a few years. The use of the halocarbons became almost universal for domestic refrigerators and smaller condensing units up to 15 or 20 hp. Air conditioning, particularly in the warmer

97

parts of the nation enjoyed steady and rapid growth. Today almost every grocery store, drive-in market, restaurant, beverage store or any food handling facility contains some refrigeration of a total tonnage that would greatly surprise the average layman.

The simplest and smallest installation consists of a single room with its own cooling coil and condensing unit. This type of installation is common to the small and medium-sized markets and many other applications requiring a limited amount of refrigerated space. It may be at either cooler or freezer temperature. Size of the condensing unit for this type of installation will usually vary from ½ or ¾ hp up to as large as 10 or 15 hp. These small plants are usually installed for complete unattended automatic control and operate normally without attention until some trouble occurs at which time a service man is called in to make repairs. The service may be performed by the company operating the storage, as in the case of a large operation with multiple stores, or by an independent service company. These installations are amazingly rugged and will operate for long periods of time without attention. Operation and efficiency is, of course, improved if regular maintenance is performed instead of waiting for something to break down before calling for service.

REFRIGERATION OF SMALL COLD AND FREEZER STORAGE PLANTS

The refrigeration cycle of the small plant usually consists of a single condensing unit which contains the refrigeration compressor and drive, condenser, receiver and some controls, all mounted on a common base. In some market installations, a common machine room containing a number of individual condensing units will be found and in the smaller installations, the condensing unit may be located outside the store at ground level or on a roof top.

Air Cooled Condensers

There are several methods of refrigerant condensing in common use today; all of which have merit. Where only 1 or 2 small units are in use, an air cooled unit condenser for each compressor will most often be used. This type of condensing is relatively trouble free and requires little maintenance. A condenser of sufficient size to give a head pressure as low as is practical is to be desired for economical operating costs. With the larger than conventional air cooled condenser, however, first cost will run somewhat high. Some compromise is usually used between optimum operating conditions and first cost that is satisfactory to the

owner. Air cooled condensers should be selected, particularly in warm climates, for a temperature difference not greater than 15°F between ambient air temperature and condensing temperature. With this selection, or a lower temperature difference, operating costs will not be excessive. In equipment rooms containing multiple air cooled condenser units, it is vital that adequate fresh air quantities be introduced into the room so that entering air temperature to the condensers will not be higher than outside ambient air temperature. Exhaust air fans may also be required depending on the design of the equipment room. When air cooled condensers are roof mounted, care should be exercised to prevent elevated air temperatures of air flowing over a hot roof surface from entering the air cooled condenser.

Some type of winter control is advisable when air cooled condensers are in use. Cold ambient air tempertatures across the condensers can cause a condensing pressure so low that expansion valves on the low side equipment will not feed properly and trouble will be experienced with the equipment during periods of cold weather. In systems using thermal expansion valves, a relatively constant pressure of the high side needs to be maintained to achieve proper refrigeration. In mild climates, constant or near constant head pressures may be maintained by the cycling of the condenser fans from a pressure control actuated by the high pressure side of the system. In colder climates, fan cycling may not be sufficient and other means must be employed. Dampers operated from a pressure control may be installed over the condenser to prevent or regulate the flow of air over the condenser surface. Another type of control is a hold back valve that maintains high-side pressure by shutting down on the outlet of the air cooled condenser, thus flooding part of the condenser with liquid refrigerant so that only enough of the surface is active to maintain the set pressure. This has the same effect as putting a smaller condenser in the circuit automatically as the ambient air becomes colder. Various methods of accomplishing this are used by different manufacturers. It is regretable that higher than normal head pressures must be maintained in cold weather, with corresponding high operating costs, to maintain proper flow characteristics of the refrigerant to the evaporator, but this is the accepted practice in a large number of the smaller installations and in some that are not so small.

Water Cooled Condensers

A second method of condensing is by the use of water cooled condensers. Water cooled condensers, when clean and properly sized, will

operate somewhat more efficiently than air cooled condensers, particularly during periods of elevated ambient air temperatures. In areas where city water is abundant and cheap, the use of city water through the condenser and to the drain gives a very efficient operation. Unfortunately such sources of condensing water are rapidly becoming nonexistent and are now usually either prohibited by law or are too expensive to consider for this use. The alternate to the use of city water is the use of a cooling tower and the recirculation of the condenser water. Cooling towers give good results when of adequate size for the load imposed on them but since they operate by evaporation of water to obtain cooling of the water, they are subject to buildup of solids from the evaporated water and the condensers are subject to scale buildup. All of this demands some type of water treatment or extensive bleedoff, or both so that maintenance and water treatment can become quite a factor in the operation of cooling towers. In areas where water is unusually bad or hard, rather expensive treatment may be required, the cost of which will tend to offset savings from improved operation.

An alternate water cooled condenser operation that is used quite extensively, particularly in multi-machined installations is the indirect system. In this system, a closed water circulating system is maintained through the machine condensers. This allows a permanent water treatment to be used in the circulating water since there is no evaporation or deterioration of this water. With the closed circuit of treated water, no scaling can form in the condensers and they will operate at peak efficiency at all times. The treated water is cooled in a closed circuit coil located in the cooling tower. The usual form of the water cooler is the use of an evaporative condenser with the condenser cooling water circulating through the tubes of the evaporative condenser rather than refrigerant. About 4 gal of condenser water per minute per ton of refrigeration is circulated through the various machine water cooled condensers by means of a circulating pump and the various machine condensers are usually somewhat oversize to offset any increase in condensing pressure due to the double heat transfer of the system. With a 78°F W.B. temperature of entering air, water temperature in the condensing system will normally be 95°F or lower. With this temperature as the hottest in the cooling coil, very little scaling will occur from the water sprayed over the coil and, except for areas of very bad water, only slight treatment of the spray water other than a good bleedoff will be required. Water from the machine condensers can also be diverted through heating coils and used as an aid to heating of occupied areas during cold weather which can result in some savings in heating costs.

This indirect cooling system has rather wide use, in some areas, in super-markets where a considerable number of condensing units are in use in a single installation. Water temperatures to the condensers from the indirect cooler are rather easily maintained where desired in winter by means of a thermostat controlling the fan of the water cooling equipment.

Evaporative Condensers

A third method of condensing is the evaporative condenser. When a number of separate machine circuits are desired, the evaporative condenser may be multi-circuited with a separate circuit for each compressor. Good operating costs and good results are obtainable from evaporative condenser operation, particularly if the condenser is selected on a generous basis. It will generally be found that oversizing of condensing equipment is a very wise investment in any plant and that dollars spent on extra condenser surface will return with friends.

Evaporators

Low side equipment for the small plant most often consists of a single finned evaporator with forced air circulation by means of a direct mounted fan of the propeller type either blowing through the coil bank or placed on the outlet side and pulling the air through the coil

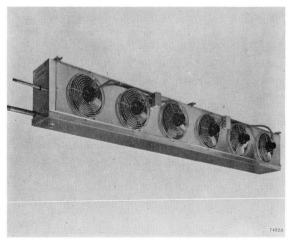

Courtesy of York Div., Borg-Warner Corp.

Fig. 8.1. Multi-fan Coil Used in Small and Inter-mediate Plants, Available in Various Types for Cooler or Freezer Operation

bank. Fin spacing will vary from about 4 to 8 fins per inch. At 35°F and lower, the wider fin spacing is desirable since at this temperature frost will tend to build at a fairly fast rate and the narrower fin spacing tends to block air circulation with the formation of frost in a relatively short period. Where room temperatures are above 35°F, more fins per inch of coil length can safely be used, thus reducing the first cost of the coil surface.

Below 35°F room temperature, some positive means of evaporator coil defrost is normally required. At 35°F, air defrost is possible by shutting off refrigeration and allowing the room air to circulate through the coil, forced by the unit fan, and melt off the accumulated ice and frost. As the room temperature approaches the freezing point, this procedure takes a longer period of time until the excess time for complete defrost may cause the room to warm up past a safe point for the merchandise stored. Also the coil and machine capacity may need to be increased to compensate for the shorter running time available after the prolonged defrost period.

Defrosting Small Coils

Forced defrosting of small coils is usually accomplished by electrical heat, by hot gas from the compressor or by water. Each system has merit and all are used at times, depending on the choice of the designer. At temperatures down to 0°F, electric defrost is favored by many designers and operators as it is the least complicated and simplest system to install. Larger electrical service is sometimes required, however, to handle the electric heater load during defrost. Hot gas defrost probably does the best job at temperatures below 0°F and lends itself well to automation. Water defrost is not too common in the small automatic plant since it is rather difficult to automate with full safety.

Refrigerant Compressors for Small Installations

There are three types of compressor units in common use for the small plant. These are the full hermetic motor compressor unit, the semihermetic motor compressor unit and the open type compressor either belt or direct driven.

The full hermetic motor compressor unit combines a compressor and a driving electric motor into a single unit with a common shaft and the entire assembly encased in a steel enclosure. There is no access to the motor or compressor except by actually cutting open the enclosure; hence this type of motor compressor unit is changed out as a complete unit after a failure. The hermetic type unit is most often the lowest in first cost. If properly installed and good installation practice followed, the hermetic unit will perform very well, particularly in the smaller sizes up to 3 or 5 hp. This type unit is widely used in small air conditioning units and also up to 2 hp or so for commercial applications. Speeds are usually 1,750 or 3,500 rpm, the higher speed coming into vogue in the past few years. The higher speed compressors are normally somewhat less efficient than the slower speed units but are usually cheaper and more compact and thus have a competitive advantage in small units where operating cost and efficiency are not prime considerations.

The semihermetic motor compressor unit is also a direct drive unit with the motor and compressor mounted on a common shaft. There is one difference, however, that in the semihermetic unit, the compressor valves may be field inspected and field repaired as well as the pistons and rods. Bolted cover plates are used in this type of unit over the compressor end so that field adjustments and repairs can be made. Semihermetics are commonly used in sizes much larger than found in full hermetic units. Commercial storage rooms are refrigerated by this type of compressor unit in many instances. The reliability of the semihermetic compressor unit is somewhat greater than the full hermetic to the extent that the compressor can be field repaired where the full hermetic cannot. The full hermetic and semihermetic compressors both present the hazard of system contamination in the case of a motor burn out. This can lead to costly delays and expensive system clean up in case of motor failure and is a determent to the use of these compressors in some instances in cold storage or other process work where down time cannot be tolerated.

The open type compressor, probably the most reliable and commonly found type of refrigeration in the medium-sized plant, consists of a compressor with a driving shaft extending out of the compressor through a seal and driven either directly or through a V-belt drive from a separate electric motor. From the reliability standpoint, this allows the electrical parts of the system to be entirely isolated from the refrigeration. If motor burn-out or failure occurs, no contamination of the refrigeration cycle results and when a new motor is installed, there is no clean up of the refrigeration system required. This type of compressor system is usually employed in the medium-sized cold storage application and almost always in the larger systems. In the belt type of drive, speed can be adjusted to suit the refrigeration load, merely by changing pulleys. The compressors usually operate at somewhat slower speeds than the hermetic and semihermetic compressors and the larger direct driven units seldom exceed 1,200 rpm. Direct driven units at this speed will be found for both ammonia and halogen refrigerants.

Unitized System of Refrigeration

The small and intermediate plant and also an extension to a larger plant will sometimes employ a unitized system of refrigeration. This system is a complete factory assembled package consisting of the complete high and low side in one package, or unit. There are actually two types of this unitized equipment. In one system, the compressor and condenser, receiver and other high side components are directly piped to a complete low side system consisting of the evaporator and fans and controls. This completely piped system is inserted in a wall opening cut in the cooler or freezer wall so that the high side equipment is on the outside of the wall and the evaporator extending in to the cooler or freezer. This system has the advantage of a complete factory fabricated system where controlled fabrication procedures and cleanliness of the system may be maintained. The system is maintained in a completely assembled state during installation and no field refrig-

Courtesy of York Div., Borg-Warner Corp.

FIG. 8.2. PLUG TYPE UNIT COMPLETE WITH LOW AND HIGH SIDE FOR INSTALLATION IN WALL OF COOLER OR FREEZER ROOM

eration piping is required. Installation is rapid and with a minimum of labor. Units are made for coolers and also for low temperature applications with automatic defrost. The disadvantage of this type of unitized equipment is that it is limited to applications where an outside wall can be utilized that is free of obstructions that might interfere with the flow of ambient air to the condenser. The second type of unitized equipment is the prefabricated system consisting of a separate high and low side and requiring the field connections between them and also defrost piping in low temperature applications. The advantage of this type system is that the high side unit may be located remotely from the evaporator unit such as a roof top mount or on the ground depending on the configuration of the complete system. This type of unit is usually made in a larger variety of sizes than the completely unitized system since a heavier highside may be used when it is on the ground or supported on top of a roof instead of balancing on a wall. The installation of this type unit requires some refrigeration piping done in the field with the attendant hazards of moisture or other contamination in the system. On systems of comparable size, this system requires more installation labor than the complete unitized system; but on the other hand, much larger systems may be utilized by this method.

Both types of unitized systems are produced almost exclusively using one of the halogen gases as a refrigerant and for air cooled operation. Refrigerants 12, 22 and 502 are all used depending on the individual manufacturer and on the application.

Intermediate Cold and Freezer Storage Plant Condensers

Careful consideration should be given to the type of condensing media used, particularly for the intermediate sized plant. The small installation with a single coil and a single machine almost universally uses an air cooled condenser. Total operating cost is not too great in this sized system regardless of the type used and the reduced maintenance cost with air cooled condensers, providing the air cooled condenser is of adequate size, will usually offset any increased power cost. In the larger plants, however, air cooled and water cooled condensers or the use of evaporative condensers should be carefully compared for first cost, operating cost and general plant efficiency. In areas where the water quality is good, it will quite often be found that the advantage lies in water cooled equipment, both from the first cost and operating and owning cost. The recent trend has been to air cooled equipment when the halogen refrigerants are used and there is no doubt that air cooled condensers definitely have a place and are to be preferred in many instances but care should be exercised to actually investigate as to which system is best for any particular installation. Too often one system or another comes in to vogue as a cureall for all conditions with no basis in fact for such an assumption.

The small single vault installation in many instances utilizes the prefabricated types of construction. Factory fabricated insulated panels in standard sizes are assembled and locked together on the job to build a vault of the required size. A variety of insulations and wall finishes are available from various manufacturers. While the most frequent type of installation utilizing this method is the small vault, installations of rather large size are made from prefabricated panels and this type of structure may well be expanded in the future to almost any size vault required. First cost is normally the prime consideration as to whether the vault or vaults will be built in place or prefabricated.

The intermediate plant may be considered in the range of 3,000 to 4,000 sq ft to as high as 30,000 to 40,000 sq ft. This classification is entirely arbitrary and is only used as a guide for the selection of systems. At the low end of this range, installations as described for the small plant are quite often used. As the size increases, more sophistication is often built into the plant and controls so that a more automated operation will result with more safeguards and more items controlled than in the smaller plant. As plant size increases, self-contained units as described previously may be used in multiple. A multiplicity of units makes for a safe plant from failure since each system is self contained and not dependent upon another system for operation. Thus, if one system breaks down, the remaining systems are still operative. While the safety factor is important in any plant, the operating cost of the air cooled, self contained package may become excessive and first cost may also run somewhat high. The package unit with roof or wall mounting does save valuable floor space but as plants become larger, however, servicing of these units becomes a problem since, in effect, there are now multiple small machine rooms, each one separated from the other. Leaks can occur in the halogen systems and go unnoticed until the refrigerant charge is gone. Usually, as larger plants are conceived, the design will favor a central machine room with multiple equipment. Two or three evaporator temperatures will usually be common. As a plant design becomes larger, ammonia refrigeration seems to be preferred. This is due to several reasons. Ammonia machines can be used in parallel operation with no difficulty. Oil return to compressors is not a problem although oil can cause low side problems. Because suction line velocities can be held to whatever limits are desirable, pressure drops can be made minimal. Direct expansion may be used in rooms at considerable distance from the machine room without suffering serious pressure loss in the suction temperature. The ammonia system will have, as a rule, a somewhat higher first cost but the equipment is usually heavier and has a longer life expectancy. A much higher degree of surveillance is required in a plant utilizing one of the halogen gases since refrigerant leakage is not normally detectable except by special testing methods. The odor of ammonia is, of course, easily detected and normally any leaks are stopped with alacrity.

Direct Expansion in Intermediate Size of Plants

The intermediate plant will quite often be designed for direct expansion refrigerant in the entire plant although circulating chilled brine is still used in some designs. The use of circulating brine chilled at a central point allows the use of almost any prime refrigerant desired, in that the refrigeration mains are short so that pressure drops offer no temperature or capacity problems. Brine is easily controlled either by flow or mixing or by-passing in the cold storage unit coolers. There are a number of good characteristics in a brine plant. All of the refrigerant is normally confined to one area which makes leak detection easier in that there is less area to patrol. This is especially valuable if halogen refrigerants are used. Ammonia still has the same advantages as previously mentioned whether used for brine chilling or direct expansion. Calcium chloride brine has been used almost exclusively through the years since it has good temperature nonfreeze characteristics where used, a low first cost and it is easily obtainable. Recently some of the glycols and other substances have been used as brines and also some of the halogens have been used in low temperature work. Calcium chloride is rather highly corrosive but with proper treatment, it can be kept neutral without much trouble. With proper neutrality and treatment, calcium chloride brine can be circulated even through copper lines although this procedure is not recommended. The prime disadvantage of chilled brine in a plant is that double heat transfer is required which impairs efficiency to some extent. Also brine circulating pumps require additional power to operate.

Evaporator selection is important in the design of any cold storage plant. As stated, the smallest rooms usually consist of a single evaporator of the finned type, with propeller fan for air circulation. As plant and room sizes increase, multiple small evaporators may be used and placed to blow so that duct work is not required. Where no duct work is involved and the fan is either directly in front or to the rear of the coil, propeller fans are most frequently used. In areas of heavy frost formation, experience has shown that the fan placed in front of the coil, drawing air through the coil and discharging it in to the room is to be preferred over the fan mounted behind the coil and blowing the air through the coil. In pulling air through the coil, the fan is less subject to frost or ice build up on the blades. Ice on a fan blade can throw a fan out of balance so that sufficient forces are set up to literally tear the fan and motor out of the mounting bracket.

In larger rooms, it is preferable to utilize multiple units of somewhat larger size than used in the small rooms. The actual size of the units is influenced by the air circulation pattern desired, head room available and other factors. Larger units, utilizing integral horsepower blower motors are subject to less maintenance ills than the small fractional horsepower motors used on the smaller units. A comparatively long throw of cold air can be accomplished in a cold storage room with several evaporators if they are set in a line, usually along one wall, all blowing out into the room. The resultant air movement is multiplied as the entire room volume of air tends to respond to the multiple fans.

Location of Evaporators

Location of units in a cold room is not extremely important. Units should be set so that the blow from the units in a room will give even coverage. Several units in parallel along one wall of a room is quite

Courtesy of York Div., Borg-Warner Corp.

FIG. 8.3. CEILING TYPE INDUSTRIAL REFRIGERATION UNIT

Available for cooler or freezer operation. Fans pull air through coil and blow out into room.

often desirable. Duct work is rarely necessary and if measurements are taken, it will normally be found that there is very little temperature variation in various parts of the room regardless of the location of the cooling units as long as there is an adequate quantity of air circulated.

Ceiling type units of a capacity of from 5 to 20 tons are used frequently in the larger cold storage rooms, both freezer and cooler, because they are out of the way and use no valuable floor space. They are also high enough off the floor that they are not subject to damage from materials handling equipment. This type of installation, however, can lead to neglect of maintenance since it is usually some trouble to get to the units and for this reason, some designers prefer floor-type units with short vertical ducts and a gooseneck connection to throw the air out at ceiling level. This installation takes up some floor space but less unit problems are likely to be encountered from the service and maintenance standpoint. Adequate guard rails should be used to prevent contact between floor units and fork-lift trucks since in any contest of this sort, the cooling unit invariably comes off second best.

Pipe coils both plain and finned are still used in some plants although

Courtesy of York Div., Borg-Warner Corp.

FIG. 8.4. FLOOR TYPE UNIT FOR INDUSTRIAL COOLING USE

Available for cooler or freezer operation. Unit is capable of operation with duct work attached to fan openings.

not many plants are being designed in this manner at the present time. Coils are costly and the roof and ceiling structure of the room must be designed with sufficient bracing to support the weight of the banks of pipe coils.

Both pipe coils and air units can be operated on either direct expansion or circulating brine. Control of brine is relatively simple and is best accomplished by a motorized valve either controlling flow or by a three-way type by-passing the coil. The latter type is probably to be preferred since it keeps full flow of brine through the chillers and pumps at all times.

Direct expansion can be controlled in a number of ways. The refrigerant liquid circulation cycle is increasingly being used where direct expansion of refrigerant in the coils is desired. Control of temperature can be accomplished by shutting off the flow of liquid refrigerant to the coil by a solenoid valve actuated from a room thermostat. A second and perhaps better method, particularly where high humidity is desired, is the modulating back pressure regulating valve controlled from room temperature. This method allows the minimum temperature difference to exist between refrigerant and room temperature.

SUMMARY FOR SMALL AND INTERMEDIATE PLANTS

The smallest rooms will usually consist of a single air cooled condensing unit with a single direct expansion coil in the cold room. Control will be off and on from a thermostat either starting and stopping the compressor or operating a liquid line solenoid valve which will allow the compressor to pump down or shut off on a pressure control. Various means of automatic defrost may be used.

As larger rooms are encountered and also a multiplicity of rooms, a number of condensing units may be employed with multiple evaporators to each unit. Also the unitized systems for wall, roof or ground mount of the condensing unit of the factory fabricated type may be used, singly or in multiple on a single room.

The larger of the intermediate plants will normally have a central machine room with machines in multiple and brine or direct expansion to the various rooms. Refrigerant for the larger plant, if direct expansion, will probably be ammonia due to the ease of handling and ability to be used with long suction lines and minimum pressure drops. Liquid recirculation will also be found in rather common use feeding multiple floor or ceiling forced air evaporators.

BIBLIOGRAPHY

AIR CONDITIONING AND REFRIGERATION INST. 1963. Standards for application and ratings of centrifugal liquid chillers, 555–63. Arlington, Va.

ASHRAE. 1965–1966. Guide and Data Book, Fundamentals and Equipment. Am. Soc. Heating, Refrig. Air-cond. Engrs., New York.

ASRE. 1957–1958. Air conditioning Refrigerating Data Book, 10th Edition. Am. Soc. Refrig. Engrs., New York.

BUSHLER, L., JR. 1951. Multi-cylinder high speed ammonia compressors. Refrig. Eng. 59, 459–463, 498–500.

CARRIER, W. H. 1926. Centrifugal compression as applied to refrigeration Refrig. Eng. 12, No. 8, 253–268.

CHURCH, A. H. 1950. Centrifugal Pumps and Blowers. John Wiley & Sons, New York.

HUNSAKER, J. C., and RIGHTMIRE, B. G. 1947. Engineering Applications of Fluid Mechanics. McGraw-Hill Book Co., New York.

INSTITUTE OF GAS TECHNOLOGY. 1957. The absorption cooling process. Bull. 14. Ill. Inst. of Technology, Chicago, Ill.

LORENTZEN, G. 1952. Influence of speed on compressor volumetric efficiency. Refrig. Eng. 60, 272–275.

MCCORMACK, A. A. 1953. Compressor design variation. Refrig. Eng. 61, 622.

OSCARSON, G. L. 1952. Application of synchronous motors to reciprocating compressors. Refrig. Eng. 60, 260–262, 306.

CHAPTER 9

Machine and System Selection for Large Refrigeration Plants

INTRODUCTION

Large cold storage systems of over 100 tons refrigeration capacity are almost always tailor-made to suit the requirements of a particular area and the desires of the owner. This chapter will cover several broad areas of design. It should be realized that no two large plants will be identical and that many variations from a norm will exist.

REFRIGERANTS FOR LARGE REFRIGERATION PLANTS

Ammonia as a prime refrigerant remains in a favored position in most large cold storage installations. The continued use of ammonia in the larger plants is justified for a number of valid reasons: parallel operation of machines with ease; absence of serious oil problems; the ability to utilize long refrigerant suction lines without undue pressure losses along with easy two-staging; high flexibility in operation; and good efficiencies. All of these are factors that tend to influence decisions in favor of ammonia. Another important factor in the use of ammonia refrigeration is familiarity with ammonia systems by many design engineers. Ammonia has been used as a refrigerant for a long period of time, and in that time, many data have been established so that systems can be designed with considerable assurance that they will operate as desired.

Unfortunately for the halogen refrigerants, the operating criteria for these plants are almost always set up with a condensing temperature some 10°F higher than that normally used for ammonia plants, and the bulk of the data used for design are based on the higher head pressure. Operating costs for the halogen gas system will show up higher, therefore, when two standard designs are submitted since the halogen refrigerant standard will, as a rule, be designed for the higher head pressure. The higher head pressure also causes higher condensing temperatures and these higher temperatures, in water cooled condensing equipment, tend to build up early scale deposits and with limited condenser surface, a snowballing effect often times requires excessive maintenance and scale removal. All of this is against the use of the halogen refrigerants. It is

112

usually true that the standard halogen refrigerant plant has a low first cost, but this is illusionary when the later maintenance costs are encountered.

OVERRATED REFRIGERANT COMPRESSOR SYSTEMS

The tremendous growth of air conditioners has seen an almost equal growth in refrigeration equipment for air cooling. Almost all equipment for air cooling has been designed for use with one of the halogen gases. Competition has been extremely keen in the air conditioning field with the result that first costs have been cut to very low levels and equipment furnished is sometimes barely large enough to perform as it should. Since first cost is most often the determining factor in air conditioning equipment, efficiency and operating costs may not be considered to any great extent in evaluation of equipment furnished. Unfortunately, some of this practice has spilled over into the industrial refrigeration field, in many instances, by designers who are not fully aware of the requirements of the industrial refrigeration plant. In other instances, manufacturers data which was never intended for use in industrial applications is used for this purpose with frequent failures.

IMPROVED DESIGN OF REFRIGERANT COMPRESSORS FOR COLD AND FREEZER STORAGE

The good designing engineer, regardless of the refrigerant used, should design on sound engineering principles so that the system he recommends will not only perform all of the functions desired by the owner, but will also perform efficiently and economically and with a minimum of costly maintenance. This type of plant will bear a higher first cost but it must be borne in mind that in a cold storage plant, equipment must operate 24 hr a day and every day in the year and that efficiency is still an item to be sought and low operating costs are to be desired.

Ammonia equipment has always been designed with heavy duty use in mind and the equipment standards, adapted long ago, were predicated on a long machinery life and low operating costs. For this reason, the cost of an ammonia plant will normally exceed the cost of other refrigerating plants.

It is very possible and desirable that in the coming years, equipment and refrigerants will come into being that will change the present concepts of refrigeration. Much work has been done to promote the halogen refrigerants with considerable success but not much effort has been expended to stimulate the building of better and more efficient compressors, in the range of 100 tons or so or to promote the obvious advantages of efficient operation in the industrial field, except by the use of equipment primarily designed for lighter duty such as used in the air conditioning field.

The foregoing is not meant as a criticism of any refrigerant, but a plea that all factors be studied and the proper refrigerant for a particular job along with its related equipment be used. The greatest use of the halogen refrigerants has been in the air conditioning and small and medium commercial field and it is only natural that concentration of effort should be in this direction. Conversely, ammonia because of its peculiar properties and safety hazards has been best used in the industrial field and hence has concentrated in this field where heavy duty machinery is usually required. There are, of course, cross overs from one field to the other and there are grey areas both ways. In very large tonnages, a great amount of progress has been made with centrifugal and screw type compressors suitable for any refrigerant.

DIRECT AND INDIRECT COLD STORAGE REFRIGERATION PLANTS

There are essentially two types of cold storage refrigeration plants. These are the direct and indirect systems. In the direct system, direct expansion coils are used in the cold rooms with the compressor and high side equipment concentrated in a central machine room. The indirect system utilizes this same type of machine room and, in addition, adds brine chillers. Brine is chilled in the machine room and is circulated by pumps through pipe lines to the cold rooms where it is circulated through the cooling units in the various rooms. Not all plants are all one type or all the other and both types are quite often found in the same plant. One system or the other, however, will usually predominate.

The normal large cold storage operation will have at least two and sometimes more refrigerant temperatures. Almost without exception there will be a refrigerant carried at a proper temperature level for cooler storage and another for freezer storage. In addition to these two basic temperatures, there may be a lower freezer temperature line when commercial freezing is accomplished and also higher than cooler temperature lines may be carried if a considerable amount of specialized higher than cooler temperatures is utilized. Dry storage space in connection with cold storage facilities may be held at temperatures of 70° to 80°F during hot summer months since this temperature storage can sometimes demand premium rates for canned goods and other products that can be damaged by extreme heat. This demands chilled water from a high suction line or direct expansion at relatively high suction temperatures.

The large machine room will therefore contain a minimum of 2 and possibly up to 4 or 5 temperature lines to the warehouse space. One of the chief advantages of the central machine room is flexibility. By cross connections and the use of two stages, it is almost impossible for a breakdown to occur that will seriously affect the overall operation of the plant. This is particularly true with a multiplicity of machines. Freezer storage is most often handled by booster compressors, particularly in ammonia plants. The boosters normally discharge into an intermediate

Courtesy of Alford Refrigerated Warehouses, Corpus Christi

Fig. 9.1. Construction Photo Showing Forming for Machine Room Floor

Entire floor is thickened for machine supports. Note electrical conduits are run to various machine and equipment locations from electrical panel location.

pressure that is also used for cooler storage. The optimum booster compressor discharge may be somewhat at variance from the suction pressure required for the coolers but usually is so close that it is impractical, from an efficiency standpoint, to carry the optimum intermediate pressure plus that required for cooler storage. If air conditioning is required for storage, or higher than normal cooler storage temperatures are required, another higher suction will sometimes be used. All high stage machines in the machine room should be valved so that any machine can operate on any of the high or intermediate pressures. Also if there are both blast or low temperature freezers and standard storage freezers, the boosters serving these freezers should be interconnected so that they can be valved into whatever suction duty is required. Variations will be found in valving of machines depending on the individual plant design.

Fig. 9.2. View Showing Poured Floor for
Machine Room and Steel Framing

Machine blocks in background and conduit stubs in foreground. Note crane rails being installed for hand crane over all components of machine room.

Design for Flexibility

By interconnecting the various machines in the machine room, great flexibility can be achieved. When extra heavy loads are encountered at one suction pressure, spare machines or machines from a lightly used suction main may be cut in to help out with the critical load. In a typical ammonia machine room for a large cold storage plant, it is desirable to operate with the high stage compressors of a size of from 100 to 200 hp each. There is a definite advantage in having all machines of the same size or in multiples of the same size. Booster compressors are desirable in the area of from 100 to 150 hp. Boosters are also desirable if all are the same size or multiples of the same size.

High stage compressors should all be interconnected as previously stated. At least one spare machine in the high stage group is desirable. This machine can be cut in to any high or intermediate suction in case of a machine failure and also when the plant is operating and receives a sudden very high demand, the spare compressor can be used to add to the plant capacity for the duration of the high load period. In keeping

Courtesy of Alford Refrigerated Warehouses, Corpus Christi

Fig. 9.3. Partially Completed Machine Room Showing Compressors on Foundation Blocks, Piping Being Erected, and With Crane in Background

the high stage compressors to horsepowers of under 200, good flexibility can be obtained under automatic control. With medium speed multi-cylindered machines with capacity controls, automatic sequence control can operate very smoothly, cutting in machines and capacity control devices as required to maintain a constant plant suction pressure.

Many cold storage operations use the medium speed multi-cylinder compressor, either direct driven or belt driven at operating speeds of 900 to 1,200 rpm. This type of machine is an excellent compromise between the older slow speed V.S.A. two-cylinder compressors and the small bore high speed compressors. The first cost is less than the slower speed machines and maintenance is low when these machines are used in a properly designed plant. Long machine life may be anticipated with these compressors. Adequate safeguards are a necessity in the modern

Courtesy of Alford Refrigerated Warehouses, Dallas

FIG. 9.4. LARGE WAREHOUSE MACHINE ROOM SHOWING SHELL AND TUBE CONDENSERS IN FOREGROUND AND HORIZONTAL COMPRESSOR UNITS IN BACKGROUND

Note looped water lines to condenser for equal friction through all condensers. Ammonia receivers (2) mounted beneath condensers.

plant with large suction accumulators and intercoolers a prime requirement to the proper operation.

Booster Refrigerant Compressors

Booster compressors most commonly used are the reciprocating-type and the rotary-type. Both render excellent service and the type used is a matter, more or less, of the owners or designers choice. In a constant suction pressure plant where the low, or booster, suction is kept at a relatively constant level, the rotary and reciprocating types of compressors will both operate efficiently. This type of service is in freezer storage rooms and in continuous freezers. Where suction pressure can vary widely, as in some batch freezing plants, the rotary compressor can cause some problems when compression ratios become too high but the reciprocating compressor is not particularly bothered by this condition. Both types of compressors are used with success in cold storage operations.

The constantly rising cost of labor and scarcity of competent personnel makes automation of the modern cold storage facility a necessity. This automation is accomplished most often by the use of electric power for driving the various components of the plant by electric motors, the electrical switching and control being somewhat easier to accomplish and more stable in operation than other types of prime movers. There are exceptions to this and other means of drive such as gas or oil engines are used as prime movers and also, in some complexes, steam may be used to drive generating equipment to supply power for the entire plant. It should always be kept in mind that the cold storage plant must operate every day and 24 hr a day. For this reason, it will be found that heavy duty equipment, although having a greater first cost, will be justified by the longer life and longer maintenance free periods. In sizing electric motors for compressor drives, the horsepower required should be checked for all possible conditions of load that may be encountered in the operation of the plant and a motor selected that will not be overloaded under any conditions that may be imposed upon it by the compressor it is driving. Most electric motors have a factor of safety to allow some overloading. This factor is usually stated on the motor name plate. It is good practice to select a motor that will meet the maximum load requirements of the compressor without utilizing any of this motor reserve. Thus the reserve will always be available strictly as a reserve. If the imposed motor load is in the reserve range, the motor may not give maximum life and will be subject to the possibility of higher maintenance costs than are normally expected of electric motors.

Condensers of Large Refrigeration Plants

The larger cold storage plants, almost without exception, will use water cooled condensing in some manner, either with a cooling tower and water cooled condensers or by means of evaporative condensers. Either method will prove satisfactory when properly installed. Condensing temperatures and their corresponding pressures should be kept to the practical minimum for long equipment life and economical operation. If evaporative condensers are used, desuperheater coils for the incoming refrigerant gas can be used to good advantage, particularly in an ammonia installation. In a typical installation where an evaporative condenser is used with an entering wet bulb air temperature of 78°F, the condensing temperature should not be higher than 95° to 96°F. With a good desuperheater coil (for ammonia) and proper bleedoff of water, very little scaling problems will normally be encountered unless area water conditions are very bad. In the case of parallel operation of evaporative condensers, proper trapping of the outlets and sufficient height of the condenser outlet above the receiver should be observed to prevent liquid backup in the condensers, thus reducing capacity drastically.

In the case of shell and tube condensers used with a cooling tower, the same condensing temperatures can be maintained as with the evaporative condenser. It should always be remembered that condenser surface is a relatively cheap investment and that the rate of return in operating and maintenance savings is high. Condensers should very definitely be larger than the minimum required to get by.

Receivers for refrigerant should be sized generously. In most plants it is probably not practical or necessary to install receiver capacity for the pump down of the entire plant. Most large plants are somewhat sectionalized so that various sections of the plant may be pumped down and adequate receiver capacity should be installed to hold the charge from the largest section. It is also good practice to use a standby receiver capable of handling a bulk truck shipment of several thousand pounds of refrigerant, since cost can be lessened by buying refrigerant in larger bulk quantities. Receivers are best operated by installing them adjacent to one another if they are to be operated in parallel. They should also have adequate equalizing lines between them.

Brine Chillers

In plants where brine is used as a secondary refrigerant, at least 2 brine temperatures will usually be maintained and each brine temperature system will require 1 or more brine chillers. In large plants, it is well to use multiple brine chillers as well as multiple brine pumps. This allows for flexibility of control as well as safety. Brine chillers can also be cross-connected between systems, if desired, for added flexibility. Brine chillers most commonly in use with ammonia systems are the horizontal shell and tube type using flooded ammonia in the shell and brine through the tubes. Chillers of about 100 tons capacity are of good size for easy and flexible operation. Pumps circulating the chilled brine should be of a capacity so that either 1 or 2 pumps can serve each brine chiller. A spare brine pump cross-connected to serve any brine system is good insurance. Brine piping to the various rooms should be large for small friction drop and economical pumping costs. Insulation of brine lines should be adequate to prevent appreciable warm up of the brine between the chiller and the room units. Brine chillers should also be adequately insulated. Safety controls for brine chillers should be adequate with high refrigerant level shut offs and also alarms.

Accumulators and Intercoolers

Large accumulators and intercoolers should be used in any ammonia plant. Suction line accumulators with liquid refrigerant return systems are very valuable insurance and also assets to economical operation. Accumulators should be large enough to keep refrigerant velocity below a point where any liquid carry over to the suction line will occur. Adequate baffling should also be built in to the accumulator to prevent splashing or turbulence of the liquid refrigerant from entering the suction line. Liquid return systems may be either powered by pressure or by liquid refrigerant pumps. There are a number of ways by which the refrigerant can be disposed of from the accumulators. Some of the systems are patented and others are not. Some systems will return the excess refrigerant directly to the plant receiver while other systems will return it to the liquid line, periodically after shutting off the main flow from the liquid receivers. Still other systems will return liquid to another low pressure vessel. Almost any scheme is satisfactory if engineered properly.

Direct Expansion Recirculators

A very excellent method of feeding refrigerant where direct expansion is used is by means of the liquid recirculation method. In this system, refrigerant liquid is fed into a low pressure receiver connected to the suction of the load being worked upon. The liquid refrigerant flashes down to the temperature corresponding to the suction pressure and the chilled liquid is then pumped to the low side units in the cold storage rooms. Instead of boiling off all of the liquid refrigerant in the low side unit as is done with flooded or expansion valve operation, an excess of liquid is fed in to the low side unit. In ammonia plants, this flow will be as high as 4 or 5 times or more of refrigerant pumped through the low side unit than is evaporated. In the case of the halogen refrigerants, slightly less liquid is normally pumped. With the increased flow of liquid, the liquid refrigerant becomes, in part, a brine flowing through the evaporator and giving very high performance by keeping the entire inner surfaces of the refrigerant tubes wet with refrigerant which increases the heat transfer ability of the tubes. The excess liquid refrigerant flows back to the low side receiver along with gas from evaporation. At the receiver, the gas separates and is pumped back to the compressor and on to the high side. The returned cold liquid drops into the liquid pool in the low side receiver and is again circulated through the system. Liquid flow may be accomplished either with mechanical pumps designed for liquid refrigerant flow or by patented pressure pumping systems in which regulated high side pressure is used to force the liquid refrigerant through the low side units. In some of these systems the chilled liquid is isolated in relatively small drums, or vessels, and the high side pressure applied to force the liquid into the system. The high side pressure is normally reduced so that no more pressure than that required to force the liquid is applied. Alternating drums can be used so that a continuous flow of liquid can be assured by allowing one drum to fill while the other is feeding. In either system, pump or pressure feed, the end result is to overfeed the low side units. Liquid recirculation has a number of advantages over either flooded operation or direct expansion. Control is simplified in that all of the refrigerant flow controls are outside of the chilled areas and at one location at the low pressure receiver. A simple solenoid valve in the liquid inlet to low side unit is sufficient to shut off the flow of liquid for temperature control or for defrost application. Other controls can be used on the low side unit if more sophisticated control is desired but the basic flow controls are concentrated at one vessel which is of some advantage in any type plant. Close temperature differences between refrigerant and room temperature can be maintained by this system. Evaporator surface is used more efficiently than with other methods of direct refrigerant cooling and a minimum amount of surface is required for good results. Since more liquid refrigerant is circulated than in the conventional refrigeration system, larger liquid and suction lines are necessary. In a large plant, the first cost of a recirculated liquid system will be very little, if any more than with any other good refrigerant cycle, direct expansion or flooded. In any recirculating system utilizing refrigerant pumps, a spare pump should always be included as insurance with each system. Any refrigerant system can successfully use a liquid recirculation system but the most commonly used refrigerant with these systems is ammonia.

Economy of Two Stage Plants

The main operating economy in two stage plants is obtained by prechilling the liquid refrigerant at the intermediate pressure before using it in the low stage evaporators. This requires the use of some type of intercooler. This intercooler also serves the function of chilling the booster discharge gas to a saturated condition. For efficient and economical operation, the liquid chilling feature should not be eliminated from an intercooler. Intercoolers should be generous in size and with some reserve for future plant growth.

Control of Ice in Refrigeration Plants in Winter

Plants operating in cold climates usually have machine rooms with adequate heating, if required, to maintain temperatures above freezing in the machine room at all times. External units, such as cooling towers or evaporative condensers, however, require some type of winter protection. Ice formations in cooling towers can produce weight and stress hazards in the tower that can cause serious damage. Forced draft towers can be protected by fan operation keyed to the pan water temperature and also by by-pass water dump lines to the tower. Cooling towers need to be watched closely in extremely cold weather to prevent freezeups.

Evaporative condensers are usually roof mounted. By virtue of the fact that they do not have large quantities of water, they are usually protected from freezing by the use of sump tanks inside the plant where temperatures are above freezing. These tanks are sized to hold the entire water charge in the evaporative condenser so that when the condenser is shut down, all of the water may drain into the interior sump tank. During operation, water temperature in the evaporative condenser sump is controlled by operating the fan from a thermostat in the sump.

Normal precautions should be taken in any plant to protect any exposed water lines from freezing during winter cold spells.

Increasing Capacity of Refrigeration Plants

A cold storage plant, even when designed as a large plant at the outset, will almost invariably grow over the years since, as in any business, to be successful some growth must always take place. It is the experience of most plants that after a few years, additions of various loads and cold spaces will present loads that the original machine room design cannot cope with. This should be recognized when a plant is designed and provisions should be made to add to the items of machine room equipment as required over the years when the imposed load increases. Accumulators and other low side vessels can be designed oversize to accommodate future loads. This does not require much additional outlay in the initial plant expense and can save a considerable amount in plant expansion. Discharge and suction refrigeration headers can also be larger than required for the initial plant for this same reason. Space should also be allocated, if possible, for additional machinery at a future date. The planning for future capacity can only be carried so far since it is impossible to look into the future as to what type of machinery might be available; and it is certainly well within the realm of possibility that entirely new concepts of food preservation and storage will prevail in the coming years. It is a good plan to build larger than required of items as described earlier that do not require a great increase in investment. If a plant can be designed to operate over a period of 10 or 15 yr without requiring major additions over the original design, this is about all that can be expected and some major redesign will probably be required in the future. It would not be economical or practical to try to design for any more adequacy than this since it is impossible to guess now what methods or what equipment will be available or desirable over the years.

Evaporator Equipment in Cold Storage Installations

Low side equipment in the various storage rooms is, at present, mostly confined to some type of forced air evaporator unit, either floor or ceiling mounted. Pipe coils, long the most used method of cooling are still used to some extent but in constantly decreasing installations, since their first cost is high and fabrication is laborious. Defrosting is also a major project.

The typical ceiling type evaporator consists of a cooling coil with fins at various spacings depending on the temperature of the room and the manufacturer of the coil. Sizes of coils will usually vary from 2 to 20 tons refrigeration capacity. Air circulation is obtained by a propeller or squirrel cage fan, either blowing through the coil or pulling air through the coil, and out into the room. The coil and fan are encased in an appropriate housing and provided with a drain pan under the coil to catch drip from condensation or defrost as the case may be. Drain pans are sometimes insulated to prevent external drip from a cold pan. Experience has shown that a number of ceiling units placed in a line and blowing out from along one wall of a cold storage room can cover wide rooms without duct work and even temperatures and uniform air flow can be maintained. The more units, and fans, in the line, the wider the room that can be air spanned. With high ceilings in a cold storage room, a distance of 100 to 150 ft may be air spanned by the blower coils along one wall of the room with the blower units evenly spaced. The multiplicity of units all blowing in the same direction tends to get the entire mass of room air circulating in a parallel pattern so that the entire room is well covered with adequate air circulation.

Care should be exercised when using ceiling type blowers that regular maintenance is maintained. These coils are up and out of the way and there is sometimes a tendency to forget about them until trouble develops. Regular inspection is vital to prevent breakdown of equipment.

In cooler and freezer storage rooms, propeller fans are most common, particularly when no duct work is involved, since the propeller fan is more efficient than the centrifugal fan when very small pressures are needed. Fans may be direct connected to the shaft of the driving motor or belt driven depending on the size and horsepower required to drive the fan. The larger units usually employ slow speed belt driven fans while the smaller units will utilize smaller higher speed direct motor mounted fans. Quite often, the smaller fan units will use a special motor for direct connection to obtain a special mounting or some other characteristic desired by the designer. It is good practice, where a number of units are used, to determine the availability of replacement motors and the cost. Sometimes these motors can get so special that it is next to impossible to replace them after a few years without high expense or remodel of units to accept an available motor. In larger units with belt drive, standard motors are normally used.

Good practice in units, particularly in freezer operation, is to use units with the fan or fans mounted to pull air through the coil and discharge it out into the room. This method of mounting also has some advantages in cooler operation. With the fan pushing air through the coil, the warmest air passes over the fan before entering the coil. In some instances, such as in a unit mounted close to an entrance door, quite heavily moisture laden air can be drawn over the fan. Frost and ice can form on the blades and the blades can be thrown out of balance. In exaggerated instances, this unbalance can be sufficient to set up enough vibration to tear the fan and motor from the mounting. Pulling air through the coil tends to freeze out excess moisture before the air reaches the fan and longer and less troublesome fan operation will result.

Floor-type units are usually somewhat larger than ceiling-type units and fewer units will normally be used per room. In most instances the floor-type unit consists of a coil and fan or fans mounted above a drain pan and all encased in a suitable housing. Centrifugal fans are normally used since air must normally be conducted up to the ceiling level of the cold room and turned to spread out in the room. This imposes some resistance and more horsepower is usually required for the floor-type unit than for a comparable ceiling unit. Air entering a floor type unit also makes a 90° turn to flow through the coil and in a standard ceiling unit passes straight through the unit without turns. The main advantage of the floor unit is ready accessibility for maintenance and repair. The disadvantages are that it takes up floor space that could otherwise be used for merchandise storage and that it, if not heavily guarded, is subject to damage from materials handling equipment. Since floor-type units are usually larger than the ceiling type, fewer are used per room and piping costs will normally be less in a total installation which will about offset the normally higher cost of the floor-type unit so that the total installation cost and equipment cost will not vary significantly regardless of the type units used.

EQUIPMENT OF SPECIALIZED FREEZER STORAGE ROOMS

Sharp freezers and blast freezers and other specialized freezer rooms will most often utilize special low side units. Commercial freezing units are available as well as built-up systems. Most warehouse operations require a general freezing arrangement for miscellaneous products rather than specialized apparatus for special purposes. The most common type of freezer used in warehouse operations is probably the blast-type. These are furnished and fabricated in any number of designs but all are basically a bank of coils with large fans delivering high quantities of air directed over the coils and over the product to be frozen by means of a suitable enclosure. This enclosure, called a freezing tunnel, will have doors at one end or along the sides depending on the way loading is accomplished. Special baffles in the tunnel direct the cold air over the product at rather high velocities. Many warehouses have several such tunnels along side of each other used alternately; some loading while others are freezing. Temperatures in tunnel freezers can be maintained over a fairly wide range, depending on the desires of those requiring freezing of their products, and may range from as high as $-10°F$ to as low as $-40°F$. One of the prime purposes served by a blast or tunnel freezer is a saving in space in the warehouse. By freezing at lowered temperatures and under blast conditions, freezing time of products is kept at a minimum and the frozen products can be palletized and placed in holding freezers without spreading, thus allowing higher product concentration in the holding freezers, with more revenue from storage. Many products frozen such as eggs, boxed or carcass meats etc. are not extremely critical as to freezing time and temperature and could normally be frozen in a standard freezer if the temperature could be maintained at about $-10°F$. Auxiliary air is usually required. The use of the blast, therefore, is helpful in a warehouse operation that is working at near capacity by conserving space and to allow more revenue from usable space. Some product advantages are also obtained by blast freezing but these are not very significant in the products normally handled by the public cold storage warehouse.

Some warehouses will also contain specialized freezing equipment for products critical to freezing time and temperature. These freezing units are usually specialized and worked out as a lease or special rated operation and the design of the freezing equipment is set up by the organization requiring the special freezing technique.

Most blast and sharp freezers will contain direct expansion coils of one type or another since low temperatures are required and the double transfer of a brine system requires undesirably low refrigerant temperatures. The liquid refrigerant recirculating system is quite efficient for

Courtesy of Krack Corp.

Fig. 9.5. Blast Type Fan-Coil Assembly for Use in Freezer Sections of Cold Storage Plants; Designed for High Air Velocity Freezing

low temperature operation and is used quite frequently. Flooded ammonia is also used in many plants. In the interest of economy and short piping, the freezers are usually located adjacent or near to the machine room so that piping is kept to a minimum. This is a necessity in the case of the halogen refrigerants since pressure drops required in a long suction line cannot be tolerated at low temperatures. In warehouses using ammonia as a refrigerant distance is not a prime consideration from an operational standpoint but from the first cost of a large suction line, the economies from a short line are apparent.

ECONOMICS OF DESIGN IN LARGE COLD AND FREEZER EQUIPMENT

It may be said that the large industrial warehouse installation is almost the direct opposite of the commercial air conditioning or small refrigeration installation. In the average air conditioning design, low first cost of equipment is of prime importance to the owner and in many cases seems to be preferred to the best efficiency and lowest operating cost. Machinery, therefore, in this field tends to be light in design with not too much reserve built in. It is built very competitively to serve in a market where competition is keen and many bids are let strictly on low price.

In the warehouse system, reliability and low operating cost resulting from efficient design and application of machinery should be the first consideration of the owner. Year round operation and maintenance of temperatures and conditions in the warehouse rooms within narrow limits must be maintained to keep satisfied customers. For this reason, reliability is paramount, which means heavy duty machines and un-overloaded motors and equipment. Operating costs are also important to the warehouse owner. These costs consist of power cost and maintenance, replacement and repair costs. For best results, all of these add up to obtaining a system containing the best and most efficient components available. While it is true that this type of installation will be more expensive than that designed strictly from competitive bidding; yet the added cost as a percentage of the total warehouse cost will be small. The best available system will not cost a prohibitive amount more than the cheaper competitive system, and the benefits of the best system are obvious.

BIBLIOGRAPHY

ANON. 1968. Standards of tubular exchanger manufacturers. Tubular Exchanger Mfr. Assoc., New York.
BADGER, W. L. 1926. Heat transfer and evaporation. Chemical Eng. Cat. Co., New York.
DIVERS, R. T., and WEST, W. J. 1964. Hermetic compressors in modern applications. ASHRAE J. *4*, 55.
GONZALEZ, R. A. 1956. How to install air cooled condensing equipment. Refrig. Eng. *64*, No. 2, 58, 95.
GREIRENGER, P. L. 1962. Handbook of Heat Transfer Media. Reinhold Publishing Corp., New York.
KRATZ, A. P., MacINTIRE, H. J., and GOULD, R. E. 1930. Heat transfer in ammonia condensers, Part III. University of Illinois Eng. Expt. Sta. Bull. *209*.

McAdams, W. H. 1954. Heat Transmission. McGraw-Hill Book Co., New York.

McCormack, A. A. 1953. Compressor design variations. Refrig. Eng. *61*, No. 6, 622–624, 680–682.

Plank, R., and Kuprianoff, J. 1935. Rotary compressors for refrigeration machines. Veriene Deutscher Ingenieure. Zeitshrift. *79*, No. 12, 363–376.

Musson, H. R. 1956. Air cooled condensers for supermarkets. Can. Refrig. and Air Cond. *22*, No. 10, 28–29, 40.

Soumerai, H. R., and Shaw, D. 1961. What is the future of the large reciprocating compressor and systems? Air Cond., Heating and Vent. *58*, No. 8, 46–47.

Vieseman, W. 1956. How, when and where to use air cooled condensers. Heating, Piping and Air Cond. *28*, No. 9, 84–87.

Wile, D. D. 1950. Evaporative condenser performance factors. Refrig. Eng. *58*, No. 1, 55–63, 88–89.

Witzig, W. F., Penny, G. W., and Cyphers, J. A. 1948. Heat transfer rates to evaporating Freon-12 in horizontal tube evaporators. Refrig. Eng. *56*, No. 2, 153–157.

Wolf, J. L. 1958. Economic evaluation of condensing methods. Heating, Piping and Air Cond. *30*, No. 8, 135–136.

Yoder, R. J., and Dodge, B. F. 1952. Heat transfer coefficients of boiling F-12. Refrig. Eng. *60*, No. 2, 156–159, 192–195.

Zumbro, F. R. 1926. Test of a vertical shell and tube type ammonia condenser. Refrig. Eng. *13*, No. 2, 46–57, 57–63, 67.

CHAPTER 10

Control Components and Physical Measuring Elements in Cold and Freezer Storages

INTRODUCTION

Temperature, humidity, fluid and heat flow, pressure, dew point, power and refrigerant conditions are significant factors in the effective operation of cold and freezer refrigerated products. This chapter will consider the fundamental components that are usually employed in the essential controls to obtain operational conditions for maximum efficiency of the cold and freezer storage warehouse plant. The chapter will deal with the available commercial Control Systems to implement the practical operation of modern cold and freezer storage warehouses.

TYPES OF INSTRUMENTS NEEDED

A refrigerating plant must be provided with sufficient instruments, if the operating engineer is to know how each part of the system is functioning. Without instruments the engineer is forced to guess at the conditions inside any particular unit and cannot operate the system efficiently or locate trouble quickly. Instruments are costly and a high degree of skill is required to maintain and repair them, hence many plants lack adequate instrumentation.

Pressure Gages

Pressure gages are always supplied with each compressor or absorption unit, but frequently these need checking after some months of usage. Unless checked they may give an entirely false picture of the true operating conditions. *Thermometers* are almost universally provided the operator for refrigerated rooms, ice tanks or brine systems, however few condensers are equipped with any means to determine the temperature of the entering and leaving stream of water or refrigerant.

In the cold storage of many foods, humidity control is equally as important as the maintenance of correct temperature. In this case the type of humidity meter in general use is the wet-bulb and dry-bulb thermometer mounted as a sling psychrometer. The depression of the wet-bulb thermometer below the dry-bulb reading is the measure of humidity; this will be explained later under *Humidity Measuring Instruments*.

Thermometers

Liquid-in-glass-type thermometers are most frequently used in refrigerating plants. The etched-stem-type has graduations engraved directly upon the glass stem and consequently cannot get out of adjustment.

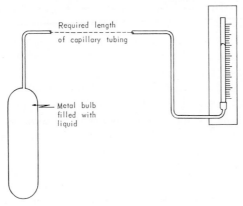

Required length
of capillary tubing

Metal bulb
filled with
liquid

Fig. 10.1. Remote Indicating Thermometer With Indicating Gage Glass

Many plant thermometers have scales attached to a protective frame or tube which encloses the glass stem, but if the scales become loose, a serious error may be incurred. All liquid-in-glass thermometers are fragile and should be protected from shocks and vibration, although it is sur-

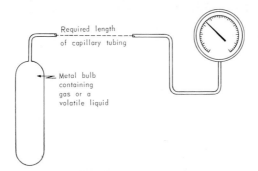

Required length
of capillary tubing

Metal bulb
containing
gas or a
volatile liquid

Fig. 10.2. Remote Indicating Thermometer
With Bourdon Gage

prising how much abuse a well-installed instrument can take and still remain in serviceable condition. Accurate observation of the indication is difficult where the light is poor or when the glass stem is dirty, and frequently thermometers are located in inaccessible places.

The limiting usefulness of a low temperature mercury thermometer is —40°F. Mercury solidifies at this low temperature. Below this temperature an alcohol filled thermometer is recommended. Precautions should be taken, however, to prevent low temperature thermometers being exposed to high temperature unless, in their manufacture, adequate expansion space for the mercury or alcohol has been provided within the instrument.

Temperature Measurement Precautions.—The precautions to take in using glass thermometers include: (a) not to exceed the upper or lower temperature ranges; (b) if the room temperature is much different from the temperature under measurement, the stem should be shielded from radiation surfaces; and (c) stem corrections may be necessary if the calibrated thermometer has been immersed at some given position in the liquid. If thermometer wells are used in pipelines and tanks, great care must be made to be sure that the liquid in the well corresponds to the temperature to be measured. Some studies on pipelines indicate thermometer well temperatures may vary 5 or 10°F from actual pipe or tank conditions. Mercury filled thermometer wells are recommended for this service.

Liquid Level Gages

Gage glasses of standard type are quite satisfactory for indicating the level of liquids which are at or above room temperature. However, they are not suitable for indicating the liquid level in a refrigerant evaporator, since the tube soon becomes covered with frost and it is impossible to distinguish the top of the liquid column. To overcome this difficulty, an insulated gage glass has been devised which consists of two concentric glass tubes sealed together and having the annular space highly evacuated and silvered, except for longitudinal strips which are left clear to permit observation of the liquid column. This method of insulation retards heat flow to such an extent that the external tube does not become cold enough to collect frost. A makeshift liquid level gage which is fairly successful, uses the formation of frost on the outside of a pipe which is connected to the evaporator in the same manner as a gage glass. The portion of the pipe which is below the liquid level is chilled by evaporation and collects a coating of frost from the moisture in the air, whereas the portion of the pipe above the liquid level is not chilled and consequently remains free from frost. The indications of this type of level gage are only approximate and response to a drop of the liquid level is very slow.

Other Temperature Measuring Instruments

There are a large number of offerings on the market of temperature measuring instruments both indicating and recording, and some for control work that indicate or record temperature differentials. The operators and managers of cold and freezer warehouses are particularly interested in mercury and alcohol, including nitrogen filled thermometers and iron-constantan thermocouples. As will be noted in Table 10.1 optical thermometers are of most value for high temperature work.

TABLE 10.1

TEMPERATURE MEASUREMENT

Instrument	Where Used	Accuracy Expected	Range, °F
Mercury thermometer	gas and liquid immersion	high	−38 to +375
Alcohol thermometer	gas and liquid immersion	high	−100 to +100
Pentane thermometer	gas and liquid immersion	high	−200 to +70
Mercury with nitrogen gas	gas and liquid immersion	high	−38 to +1000
Gas thermometer	standardization	very high	−459 to +1000
Platinum-rhodium thermocouple	high temperature and for calibration	high	500 to 3000
Iron-constantan thermometer	general plant testing	fair	100 to 2200
Copper-constantan thermocouple	general plant testing	fair	100 to 600
Chromel-alumel thermocouple	general plant testing	fair	100 to 2200
Bi-metal thermometer	for rugged controls	limited	0 to 1000
Optical pyrometer	for spot high temp radiation	1%	+4000
Seger cones	ceramic industry	3%	1000 to 3600
Boiling and melting points pure materials	standardization	high precision	Established melting points of pure material
Beckman thermometer	differential temperatures	high	10° differentials
Tempil sticks	surface temperature	1%	125 to 900
Radiation pyrometer	high temperature radiation	5%	100-up

Thermocouples

Thermocouples can be made up of commercially available thermocouple wire by welding the two dissimilar leads together. These can be calibrated on any millivoltmeter by using the melting points of pure salt, pure lead, tin and aluminum checkpoints.

In using thermocouples, care should be exercised that the couple is actually in contact with the point to be measured.

Bi-metal thermometers can be made up from available commercial bi-metal strips that expand more on one surface than the other as the temperature rises. This can be useful in snap action controls or as cutoffs at predetermined temperature levels.

Humidity Measuring Instruments

As previously mentioned, the "sling psychrometer" for measuring relative humidity of air comprises a pair of thermometers, one of which is provided with a loosely knit fabric cover upon the bulb, which is wetted with water before use. Both thermometers are mounted in a frame equipped with a pivoted handle so that they may be rapidly revolved or "slung" in the air. In air which is not saturated, the evaporation from the wet fabric cover causes the wet-bulb thermometer to read a lower temperature than the instrument having the dry bulb. The indications of the two thermometers may be used to determine the relative humidity with the aid of a psychrometric chart. A variation of this device comprises wet-bulb and dry-bulb thermometers mounted upon a stationary frame. To secure accurate results, the air should be blown over the thermometers with a velocity of approximately 1,000 ft per min, but for many applications the indication of the instrument in still air is sufficient.

A second type of psychrometer employs the change in dimensions of fibers or membranes which absorb water from moist air. One type of instrument consists of a band of approximately 50 human hairs, one end being fixed and the other end attached to an indicating pointer. In dry air, the hair loses moisture and shrinks, whereas in moist air the opposite effect takes place. Another type utilizes the twisting of a spiral element which is sensitive to humidity changes. These instruments must be calibrated at known air humidities which may be determined by the dew point method or by means of the wet-bulb and dry-bulb psychrometer. Corrections must be made for wide temperature variation, but over a limited range the relative humidity indication is approximately accurate.

Dew Point Hygrometers.—Dew point hygrometers are available at prices from a few dollars to two or three thousand dollars each. The dew point hygrometers give the value desired in a single reading rather than requiring the use of wet and dry bulb thermometers which require the interpretation of the dew point by the psychrometric chart or from tables.

Simpler dew point hygrometers utilize the phenomenon of visible condensation appearing on a bright surface such as metallic mirror that has been cooled by a refrigerant at the back. Effective use of this type of dew point hygrometer requires considerable practice to eliminate the human equation error.

Fluid Flow Meters and Indicators

Flow meters of the disc type, similar to the well-known water meter, are used to measure the flow of liquid refrigerants. For such service, special materials which are resistant to the liquid must be used and the parts must be sufficiently strong to resist working pressures. Orifice meters (see Fig. 10.3) are satisfactory for the measurement of either

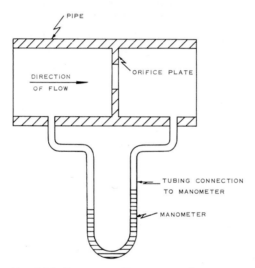

Fig. 10.3. Schematic Diagram of Indicating
Orifice Meter

liquid or gas flow and may be obtained in types which indicate the flow rate or record the total flow over a period of time. A special type of orifice meter consists of a vertically mounted tapered transparent tube through which the fluid to be metered flows in an upward direction. A specially formed plummet in the tube is forced upward until the pressure drop, through the variable orifice formed between the conical inside wall of the tube and the plummet, is balanced by the weight of the plummet. A scale beside the tube indicates the flow rate corresponding to various positions of the plummet (see Fig. 10.4).

Flow indicators are very convenient in many locations in the refrigerating plant, particularly where the flow rate is quite small. One simple type is a cell provided with windows to observe the position of a pivoted vane which is deflected by the stream of moving fluid. A second type employs a propeller or wheel which is mounted in a transparent cell and is rotated by the fluid stream. The indication of either type is approximate, but they show at a glance any great deviation from normal operating conditions.

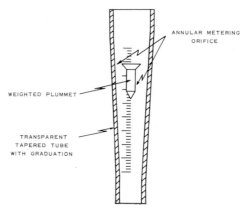

FIG. 10.4. VARIABLE ORIFICE METER

Pressure Gages

Pressure gages of the Bourdon type tubes are almost universally used. In operation, these instruments are subjected to vibration through the piping and the pressure compressors, hence their accurate adjustments may be lost and erroneous pressures indicated. For this reason a standard test gage should be provided to be used only in calibrating pressure gages which are in service. A mercury-filled manometer is sometimes employed to indicate the pressure in the low side of the refrigeration system. Its advantage is that a false indication is not possible unless oil or liquid collects in the tube above the mercury, but this condition is at once apparent and a correction is easily applied. A stop valve must be included in the manometer connection, which is closed when pressure beyond the range of the instrument is anticipated.

AUTOMATIC CONTROL COMPONENTS

All controls of a refrigeration system regulate directly or indirectly the flow of fluids, which may be liquids, vapors or gases. Hence, the majority of controls are valves; though electrical switches are also employed to start and stop compressors and thus to indirectly regulate the flow or refrigerants.

Power elements for the automatic control of refrigeration machinery may be divided into three general types: (1) elements which respond to variations of liquid level; (2) types which are sensitive to temperature change; and (3) types which are responsive to pressure variation. Some of these may directly operate the control devices, such as valves or electrical switches, whereas others actuate pilots or relays to supply energy for the operation of the control devices. For the sake of simplicity and low initial cost, it is desirable to use a power element which will directly operate the valve or switch, but large control devices may require more force than is exerted by an element of economical size.

A float exemplifies the most widely employed power element which responds to variations of liquid level. The float may consist of a hollow metal vessel of sufficient strength to withstand the pressures which are encountered in operation or during idle periods and it may be counterbalanced by a weight or spring system. When large pressures may be attained, the hollow float may be replaced by a solid metal mass which is counterbalanced by an equal mass located above the highest liquid level. When the liquid is a conductor of electricity its rise or fall may be employed to close or open a circuit between two stationary electrodes in the containing vessel.

Temperature sensitive elements depend upon the change of dimensions which occur when some suitable material is heated or cooled. The bimetallic element consists of a structure built of two metals having different coefficients of thermal expansion. A strip made by welding together two layers of dissimilar metal has the property of warping when the temperature is changed. By fixing one end of the strip, attaching an electrical contact point to the free end and providing a second fixed contact, the warping of the strip will soon open or close an auxiliary power circuit. Another type utilizes a triangle composed of 3 pivoted links, having 2 sides of 1 metal and the third side of another metal. When one link is fixed, any change in temperature causes a motion of the apex of the opposite angle, which may be used to operate control devices.

The thermal expansion of fluids may supply the energy to operate control devices. A simple form of thermally actuated electrical switch thermometer has contact points inserted through the walls of the stem in such a manner that the circuit is closed when the mercury thread touches the upper contact. In another form, the expanding fluid is allowed to act against a flexible member, such as a diaphragm, a bellows or a Bourdon tube, and thus do mechanical work, which is transmitted to the control device. The force exerted by an expanding liquid is very large, but the degree of movement is small because of the low coefficient of thermal expansion which characterizes the liquid state. Gases possess a much larger coefficient of expansion, but the greatest change in volume occurs when a liquid passes to the vapor state. Pressure sensitive power elements consist of a cylinder and piston or a flexible member, such as a diaphragm, a bellows or a Bourdon tube against which the pressure may act and cause movement of the control device. A spring system or weight is provided, which absorbs part of the energy when an increase of pressure occurs, and returns the control device when the pressure changes.

Liquid Refrigerant Control Elements

The expansion valve which separates the high side and the low side of the system and meters liquid refrigerant to the evaporator, may be considered as the most important control in the system. Manually operated needle valves are the simplest form and are satisfactory when the load is constant and the operator is vigilant. The frequent adjustment of the expansion valve which is necessary with rapidly fluctuating loads, has caused the virtual abandonment of the hand-operated valve and the adoption of automatic types. The low-side float control is used with flooded evaporators to maintain the liquid at a predetermined level. Refrigerant is supplied only when vaporization is occurring and the accurate metering effect allows any number of evaporators to be used in a system. It comprises a valve which is operated by means of a float, located either inside the evaporator or in a float chamber which is connected in such a manner that the same liquid level is maintained in the chamber and the evaporator.

In the smaller sizes, a needle valve alone is employed, but the larger sizes include a calibrated orifice in series with the valve. Division of the pressure drop into two stages reduces erosion of the needle and valve seat, which may be very rapid when throttling is accomplished in single stage. The orifice is replaceable and its bore should be of a caliber which will pass only the quantity of liquid refrigerant required for the maximum cooling load. Some low-side float controls are equipped with "snap action" valves, which are either wide open or completely closed, thus concentrating the throttling effect at the replaceable orifice and eliminating erosion of the valve parts. While this arrangement greatly prolongs the life of the valve, it may introduce an irregular action on the evaporation because of the fluctuation of the liquid level. A low-side float switch may be used to control a magnetically operated liquid valve which provides the "snap action." In a system which contains only a single evaporator, the high-side float gives effective control of the liquid supply for flooded operation. Unlike the low-side float, which accurately meters liquid as vaporization occurs, the high-side float control cannot be used to supply multiple evaporators with liquid. It is essentially a float trap which drains liquid from the condenser to the evaporator but prevents the flow of uncondensed refrigerant vapor. The amount of refrigerant in the system must be carefully adjusted to secure satisfactory operation since with this type of control all liquid is stored in the evaporator.

Too much refrigerant will cause carry over of liquid from the evaporator to the compressor while too little will not permit the system to work at full capacity.

The constant pressure expansion valve comprises a throttle valve which is controlled by a diaphragm or a bellows actuated by the pressure in the evaporator. It can be used only to control dry expansion coils under fairly constant load conditions. If a heavy load is suddenly applied, the evaporator pressure rises and the constant pressure valve cuts off the supply of liquid just when it is most needed. Conversely, when the load is very light the evaporator pressure falls, but the flow of liquid is increased when it should be decreased.

The "feeler" bulb of a thermostatic expansion valve is mounted on the evaporator suction line in a manner assuring good thermal contact and thus its temperature follows closely that of the suction vapor. When the temperature of the suction line exceeds a predetermined value, the pressure exerted by the volatile liquid in the bulb moves the bellows or diaphragm against the combined effect of evaporator pressure and the spring, to open the valve more widely. As more liquid is admitted to the evaporator, the temperature of the suction vapor is decreased and the valve opening is decreased in proportion to the temperature drop. Thus, by merely adjusting the spring tension, the valve may be set to maintain a constant suction superheat which is largely independent of the evaporator pressure.

When used with coil evaporators of great length, in which considerable frictional pressure drop occurs, it is necessary to use an "equalizer" connection between the suction line and that portion of the valve body which houses the bellows or diaphragm. This connection insures that the internal pressure acting to close the valve will be the true suction pressure at the point where the "feeler" bulb is placed.

Thermostatic expansion valves employ a temperature sensitive power element to regulate the position of the valve which admits liquid refrigerant to the evaporator. One type, which is made by a number of manufacturers, is operated by means of a metallic bellows or a diaphragm

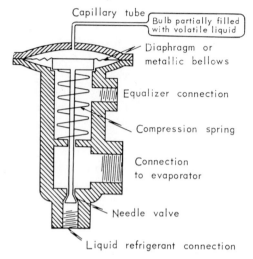

Capillary tube

Bulb partially filled with volatile liquid

Diaphragm or metallic bellows

Equalizer connection

Compression spring

Connection to evaporator

Needle valve

Liquid refrigerant connection

FIG. 10.5. THERMOSTATIC EXPANSION VALVE

which is located in the low pressure portion of the valve body and is thus subjected on one side to the pressure prevailing in the evaporator (see Fig. 10.5). An adjustable spring adds its force to the action of the evaporator pressure and tends to close the valve. The opposite side of the bellows or diaphragm is enclosed in a fluid tight housing which is connected by means of a capillary tube with a bulb containing a charge of volatile liquid.

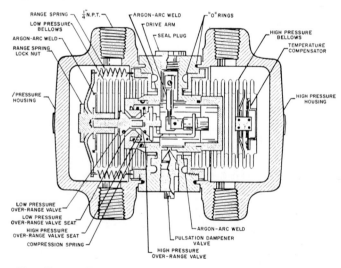

FIG. 10.6. A DIFFERENTIAL PRESSURE BELLOWS TYPE METER

Capillary Tubes

A device which functions as the expansion valve of a refrigerating machine is the capillary tube used in many domestic refrigerators and small hermetically sealed units. It reduces the pressure between the condenser and evaporator by means of fluid friction in a section of tubing having a very small bore. The satisfactory operation of this device depends upon a nice balance between compressor capacity, tube dimensions and the amount of refrigeration in the system. It is apparent that the refrigerator temperature may be controlled only by "on and off" operation of the compressor, since the volume of liquid refrigerant flowing through the tube is dependent only on pressure drop. When the compressor is idle, pressure throughout the system is equalized and thus the starting load consists only of friction and acceleration work. As the machine runs, gas is forced into the condenser faster than it passes through the capillary, thus the pressure rises until condensation takes place and liquid passes to the evaporator, which is now at low pressure. Advantages of the capillary are its extreme simplicity and the fact that there is no moving part to get out of order, while the equalization of pressure during the off-period permits the use of low starting torque motors. Strict control must be exercised in assembly to avoid the presence of dirt particles which might choke the tube.

The *constant pressure* valve maintains satisfactory temperature control when the refrigerator load is constant, but with a variable load other methods of control must be employed. A thermostatically operated valve may be used to throttle the flow of vapor from the evaporator and adjust the evaporation rate in accordance with the imposed load. In this case the temperature of the evaporator is not held constant but decreased in proportion to the demand for refrigeration. For this service the valve may be operated by a temperature sensitive power element consisting of a metallic bellows or diaphragm, a spring and a small reservoir of volatile liquid which is maintained at refrigerator temperature. Atmospheric pressure and the spring act upon one side of the bellows or the diaphragm and the vapor pressure of the volatile liquid. This unbalance compresses the spring and moves the valve to a position which permits more rapid flow of vapor and consequently an increased evaporation rate to prevent further temperature rise.

Evaporator pressure control is a simple method to control a refrigerator when the compressor must run continuously to operate other cooling equipment or to maintain an unvarying temperature level. However, maximum operation of any throttling process causes losses of energy and reduces the effective capacity of the machine. A variation of temperature is usually permissible in the operation of refrigerated devices and this allows control between specified limits by means of an "on and off" operation of the compressor. A thermal element in the refrigerated space starts the compressor at some predetermined maximum temperature and stops it when the desired minimum is reached. Thus the machine runs at approximately full design capacity and its maximum efficiency is assured. This method is particularly well adapted to the control of small electrically driven units. In certain designs the evaporator and its refrigerant contents serve as part of the thermal element which starts and stops the compressor. When the machine stops, the evaporator gains heat and within a short time the internal pressure rises to that of saturated vapor at the prevailing refrigerator temperature. Any increase of the refrigerator temperature causes a corresponding rise in vapor pressure, which is transmitted through the suction line and closes a switch. As the compressor operates the evaporator temperature drops to some predetermined value at which corresponding reduction of vapor pressure allows a spring to open the switch.

Fluid Flow Measurement

Manometers are of long standing usage especially in laboratories. If connections are tight and pressure lines free of trapped air or overflow liquid the accuracy of measurement is excellent. Many different patented devices have been added to the higher pressure manometer instruments to make them more useful for higher pressures and for measuring fluids of more than one fluid phase such as wet steam, wet gas and similar two-phase liquids.

Bourdon coils or tubes can be used very successfully for differential pressure measurements. There are several commercial designs of so-called d/p cells that utilize the Bourdon principle for the pressure elements. For their use with Venturi tubes, orifice plate and similar differential pressure primary elements care must be taken to install the instrument so that there is little temperature change in the pressurizing fluid and if the fluid in question is water that the entire meter assembly does not freeze up in winter weather.

Bellows Differential Pressure Instrument Elements

These may be either indicating or recording units. Metallic bellows type differential pressure meters to replace the earlier type of mercury manometers are rapidly invading the market of these older-type instruments. For some services 75% of the mercury manometer instruments have been replaced by bellows type pressure differential meters. This change is primarily due to the greater ruggedness, accuracy and freedom of repairs of bellows type units (see Fig. 10.6).

Primary Elements for Fluid Flow

The primary elements for measuring fluids are of early Roman origin. The manometer flow nozzle and the differential pressure bellows are typical of the instruments that are utilized to measure the pressure drop through the Venturi tube or across the orifice. This pressure drop is translated into fluid flow by a precision design combination of instrument parts with warranty accuracies of 0.5% of actual discharge.

The Modern Venturi Tube

The proportions of Venturi tubes used for metering liquids or gases are substantially the same as those originally adopted in 1887 by its inventor, Clemens Herschel. A typical form of construction is illustrated in Fig. 10.7 which shows three types of primary elements commonly used to measure differential pressures. The Venturi tube usually offers the most accurate measure of the differential pressure elements. Starting at the upstream flange, the first portion is a short cylindrical inlet which is a continuation of the upstream pipeline. This part is either machined

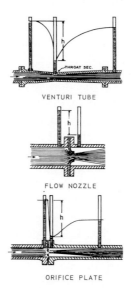

VENTURI TUBE

FLOW NOZZLE

ORIFICE PLATE

FIG. 10.7. THREE TYPES OF PRIMARY
ELEMENTS USED TO OBTAIN DIFFER-
ENTIAL PRESSURE MEASUREMENTS

inside or cast smooth so that its diameter can be accurately determined. The static pressure of the fluid at the inlet may be obtained through a single side-wall hole, or preferably several holes evenly spaced around the inlet section which lead into a piezometer ring, and from which a single pressure connection will be made.

Flow Nozzles

Briefly, the flow nozzle might be described as a short cylinder, one end of which is flared to form a flange that can be clamped between pipe flanges. This flared end forms a curved entrance leading smoothly into the cylindrical section called the throat.

The purpose of the curved entrance is to lead the fluid smoothly to the throat or measuring section so that it shall issue from the throat

TABLE 10.2
MEASUREMENT OF DIFFERENTIAL PRESSURES

Instrument	Where Used	Range	Accuracy to Anticipate % Error
Manometer	low and medium differential pressures	0 to 100 in. Hg	within ½
Bourdon coils	low pressure and high pressure	0 to 2000	within ½
Draft gages	low pressure	0 to 10 in. H_2O	within ½
Bellows differential pressure units	low to very high pressures	0 to 3000	within ½

without contraction as a straight cylindrical jet having the same diameter as the throat. The flow nozzle thus performs the same function as the entrance cone and the throat of the Venturi. It decreases the cross section of the stream in the pipe in a known ratio and also produces a difference of pressure between the entrance and the throat by accelerating the fluid. In effect the flow nozzle is a Venturi tube that has been simplified and shortened by omitting the long diffuser on the outlet side.

The Orifice Design of Today.—The orifice in its different forms is the oldest device for measuring or regulating the flow of fluids. The thin plate of sharp edges orifice has been commonly employed for many decades to measure the rates of discharge of fluids from one large tank into another or into the open air. Within the last five decades the behavior of the orifice as a device for measuring the flow of fluids in pipelines has been systematically studied and improved.

In some cases, the inlet edge of the orifice is rounded off, thus, in effect, producing a very short nozzle. Since round-edged orifices are difficult to specify and reproduce, their use has been restricted almost entirely to experimental installations or special instruments. If rounding off the edge of an orifice is necessary, it should be well rounded because experiments show that a very slight rounding off produces a decided change in the character of the flow through the orifice.

Air Current Measurement

There are many conditions where a Weather Bureau type of anemometer cannot be used to determine airstream flow rates. For this type of service the Kata thermometer may be useful. This is an instrument built like a thermometer but used primarily for air flow studies. The Kata is an alcohol thermometer with a very large bulb and with a high scale near the top divided in 125° to 130°F markings, and another set of lower temperature markings from 95° to 100°F proportionately down on the same stem.

To operate this instrument, the alcohol bulb is heated in a bath above the higher scale, then dried and placed in the air stream to be measured. The time in seconds is now noted for the drop in temperature from the upper or high scale to the lower. A chart is furnished to indicate what this time period represents in air velocity.

Heat Conductance Through Walls

The k values of unit heat transfer of wall and insulation materials can be accurately determined by the available testing laboratories and each manufacturer can supply these values for his own product. It must be observed, however, that these values are for standard test conditions in a prescribed hot box or on a "hot plate." When these same insulating materials are built into a wall they are subject to moisture infiltration, ventilation currents and radiant rays and these may effect much different actual values than the calculations predicted from the laboratory k value.

Wall conductance meters are available that will indicate whether the heat is traveling in or out of the wall, or under certain conditions moving both outward and inward from the center of the mass to the two outer surfaces. A heat flow meter can be obtained that measures heat flow in or out of a wall made up of standardized thermal resistance plate that has thermocouples attached to both faces of the plate. In operation, this plate is secured firmly to the wall under test. The temperature difference as a function of the heat flow rate as indicated by the thermocouple on each side of the panel of known thermal resistance indicates the direction and the amount of the heat flow. For accuracy this type of heat flow meter should be calibrated periodically.

Sound Measurement

The sound level meter is a desirable instrument especially for engineers of hotel chains, auditorium managers and maintenance engineers and technicians of schools and other public buildings. However, the sound nuisance is more than a sound level phenomena since the frequency of the vibrations are also of importance.

BIBLIOGRAPHY

DICK, J. B. 1950. Measurement of ventilation using tracer gas techniques. Heating, Piping and Air Cond. 22, No. 5, 131–137.

FRANK, W. G. 1932. Size and characteristics of air borne impurities. Heating, Piping and Air Cond. 4, No. 1, 35–39.

GIER, J. T., and DUNKLE, R. V. 1954. Using the heat flow meter to study heat transfer. Refrig. Eng. 62, No. 10, 63–69, and 90–96.

GROSS, E. E., JR. 1957. Noise measuring and sound control. Refrig. Eng. 68, No. 5, 49–53, 90, 92, and 94.

HOOPER, F. C., and LEPPER, F. R. 1950. Transient heat flow apparatus for the determination of thermal conductivity. ASHVE Trans. 56, No. 1395, 309–324.

HUEBSCHER, R. G., SCHUTRUM, L. F., and PARMELEE, G. V. 1952. A Low-inertia, low-resistance, heat flow meter. ASHVE Trans. 58, No. 1453, 275–286.

LYLE, O. 1947. Efficient use of steam, His Majesty's Stationery Serv., London.

NOTTAGE, H. B. 1950. A simple heated-thermocouple anemometer. ASHVE Trans. 56, No. 1402, 431–440.

PETERSON, A. P., and BERANEK, L. L. 1954. Handbook of Noise Control. General Radio Co., Cambridge, Mass.

SPINK, L. K. 1958. Principles and practice of flowmeter engineering. Foxboro Co., Foxboro, Mass.

SUTTON, D. J., and McNEIL, P. E., JR. 1954. A two-sphere radiometer. ASHVE Trans. 60, No. 1506, 297–308.

TUVE, O. L., WRIGHT, D. J., JR., and SIEGEL, L. J. 1939. The use of air velocity meters. ASHVE Trans. 45, No. 1140, 645–662.

WILLIS, A. H., and DEGLER, A. E. 1939. Air conditioning for the relief of cedar pollen hay fever, Eng. Res. Bull. 31, Bur. of Eng. Res., Univ. of Texas, Austin, Texas.

WOOLRICH, W. R. 1966. Handbook of Refrigerating Engineering, 4th Edition, Vol. 2. Avi Publishing Co., Westport, Conn.

CHAPTER 11

Control Systems for Instrumentation of Warehouses

WIDESPREAD USE OF CONTROLS IN REFRIGERATED WAREHOUSES

It is safe to say that every refrigeration plant has some type of control used to aid in the operation of the system. The control used in small systems is often quite simple. As systems grow larger, the control becomes more sophisticated and in the larger warehouse refrigeration systems, the control system can often represent a substantial percentage of the total cost.

FULLY AUTOMATED PLANT

Over the years, and very rapidly in the past few years, the trend has been to completely automated plants. Reliability of controls has improved vastly and the technique of application has kept pace so that the present day automatic plant operates smoothly and efficiently with a minimum of operational difficulty. Automation of refrigeration plants has been forced at this rapid pace due to shortages of competent operating personnel. There are simply not enough trained operators available to operate a manual plant 24 hr a day.

Plant Regulation by Automation

Scarcity of top flight operating engineers has proved quite a stimulus to the design of fully automatic plants, regardless of the type refrigerant used. By automation, a plant may be constructed that will operate reliably and efficiently if local regulations permit, with one daytime shift of operators. With automatic control of machine operation as well as room temperatures, part of the daytime crew are freed for routine maintenance and overhaul and other duties vital to good operation. It should be stressed that an automatic plant does not mean an unattended plant when the plant is large with multiple rooms and machines. Thus, the large fully automatic plant, and also the larger intermediate plant, needs a chief engineer of the highest caliber who has a thorough knowledge, not only of the plant equipment, but of the intent of the designer and expected sequence of operation of the control system. Other personnel required are determined by the size of the plant and the responsibilities placed with the machine room for general maintenance.

The savings in an automatic plant are in the fact that only 1 full shift is usually required instead of 3, but the single shift personnel needs to be of the highest possible type.

The small and intermediate plants usually operate automatically with little or no supervision. This is particularly true in the plant employing a single compressor—single coil type of installation. Controls of such a plant are very simple. Regular inspection should be maintained, either by plant personnel or an outside service organization.

Solid State Controls

Solid state controls are in rather wide use at the present time as well as other new types. It is probably a safe assumption that, in time, most controls will utilize some form of solid state circuitry. Functionally, controls should remain the same regardless of type as long as the present forms of refrigeration are used. One of the drawbacks to solid state and other unitized controls is lack of personnel who understand them sufficiently to service and adjust them to the required plant functions. No matter what type of control is used, certain functions are required and certain results are desired. Discussion of the ideal control situation in any plant is basically a description of actions desired of a control system and are valid for any type of control so long as it performs the desired function and can be coordinated with other associated control components in the same network to produce a completely operable control system.

Preferred Control Systems

The ideal control system serves several separate functions, all integrated into one complete interdependent system. For the very small plant, this system may be extremely simple and for a large complete warehouse system, it may be extremely complex. The controls should operate the complete refrigeration plant without manual assistance from the operators. It should maintain proper set temperatures in the various cold storage rooms; energize various pieces of plant equipment as needed and deenergize them as load decreases; perform automatically such defrost operations as required in various cold rooms to maintain low-side coils in good operating condition. In case of complete shut down from an external cause, it is very often desirable for the automatic control system to start up the plant in proper sequence and to bring it into full operation automatically. Temporary malfunctions, such as surge slop over of refrigerant, should be disposed of by the control system without plant shut down. The control system also must sense malfunctions that cannot be corrected by routine control methods and be able to shut down specific components of the plant, or the entire plant, if the malfunction is serious enough to warrant the shut down. In addition, if more sophistication is desired, the control system can be made to indicate failure causes, to indicate operating equipment and in cases of failure to send an alarm to an outside source.

Safety Controls

In the completely automated plant, a number of safety controls are usually employed that can be termed prime safety controls. These controls are designed to shut down the entire plant on the failure of certain components that could jeopardize the machinery in the plant. Examples of prime safety controls are high head pressure cutouts, brine

Courtesy of Alford Refrigerated Warehouses, Corpus Christi

FIG. 11.1. MACHINE SAFETY CONTROLS MOUNTED AT COMPRESSOR LOCATION

pressure failure switches, low brine temperature cutouts, low refrigerant pressure controls (other than operating control), high accumulator liquid level controls and other cutouts that will act in case of a dangerous situation developing that could cause harm to the various components of a plant. For safe operation, it is good practice that these controls be of a manual reset type since their action is usually caused by a malfunction of the plant that could cause equipment failure, if allowed to continue, or if the plant were to restart after shut down. If the plant is without attendants for certain periods of time, then the prime failure switches may be wired to actuate a signal or a telephone so that the alarm can be picked up by a watch service or by phone to plant personnel so that the plant will be restarted manually after a proper inspection is made. By manually starting the plant after a prime safety failure, potential damage can be prevented. It will be found in the well maintained automatic plant that failures in the prime safety circuits are rare and call backs to the plant for this type of trouble do not occur very often.

A second set of safety controls is also important in the automated plant. These safety controls are set to protect various units in the plant. They do not act to shut down the entire plant but only a single machine or other item of equipment. These controls may be individual machine high pressure cutouts, cylinder thermostats, bearing thermostats, oil pressure failure switches and low pressure cutouts. Vibration controls are also used on occasion to stop machinery if vibration becomes excessive. These individual safety controls can be made to lock out the particular machine they protect or can automatically reset as required in the over all design of the particular control system involved.

Operating Controls

A third category of controls are the operating controls. These are the controls that operate the various items of equipment, cutting machinery in and out to suit the requirements of the plant operation. These controls usually consist of two different sets, one set for controlling machine room operation and another independent set for controlling conditions in the cold storage rooms.

The machine room controls govern the operation of the refrigeration equipment in the machine room. Generally, it will be found that the refrigerating machines will operate on pressure controls and not directly from temperatures in the cold rooms. In this way, the refrigeration plant will maintain certain set system pressures throughout the plant independent of the number of rooms calling for refrigeration. This makes for a very flexible method of control that is relatively simple in operation and will function with a minimum of operational difficulties.

Sequential Controls

Sequencing controls are very effective for machine room control. This type of control is normally actuated by plant suction pressures. As pressure in the plant suction rises beyond the proper operating limits, the pressure control actuates to bring another machine into operation, or if capacity controls are used on the compressors to add to the capacity of the plant by manipulation of the capacity controls plus the starting of additional compressors. This controller will most often be found in the form of a modulating control motor with various control switches or cams mounted to the shaft and designed to add capacity as the shaft rotates in one direction or diminish capacity as it rotates in the opposite direction. The operating pressure control is normally of the modulating type with standard bridge circuit actuating the controller motor a few degrees at a time, up or down, as plant demands vary. In large plants with a high mutiplicity of capacity, two or more such devices may be used to cover the number of steps required. An end switch on the first controller actuates the second controller, etc.

A further refinement in the sequence control is a forcing relay. In case of plant shut down due to power failure or other external cause, the entire control sequence will automatically return to a zero position before any equipment is started up and then the plant will start up in proper sequence, bringing on one step of capacity at a time so that large connected loads do not start up simultaneously.

As the sequence controller brings in machine capacity, the various pumps and condenser fans and other related equipment are started up as required. With this type of control, the machine room can be started up from a complete shut down, all in proper sequence, starting each machine and its related equipment as required by the plant load. The control will vary the plant at any point required between full capacity and complete shut down. In most plants, more than one suction temperature is required. Separate sequencing controls can be used to control the machines assigned to each suction pressure and also for the control of booster compressors in two-stage systems. As a rule in booster compressor controls, an intermediate suction high stage machine must be in operation before a booster compressor can start. This will prevent a booster cycling in and out on safety controls and possible trouble.

Temperature and Frost Controls

The final set of operating controls are the room temperature controllers located in the various cold storage rooms and controlling temperature in the individual rooms as well as humidity and defrost as required. A cooler room above 35°F is simple to control for temperature and defrost since the defrost is normally accomplished by shutting off the refrigeration for a specified length of time and allowing room air to defrost the coil. Only a simple time clock operation is required for defrost. Temperature in the room can be controlled by means of a thermostat operating a solenoid valve for rooms employing direct expansion or liquid recirculation of refrigerant. This type of temperature control can hold a room within reasonable limits of temperature. The same type of temperature control can also be used in freezers but defrost controls are somewhat more elaborate since some positive means of defrost must be employed.

A second and very fine method of temperature control used in con-

Courtesy of Refrigerating Specialties Co.

FIG. 11.2. ELECTRICALLY MODU-
LATED BACK PRESSURE REGULAT-
ING VALVE

Electric motor operates cam in response to thermostat to regulate refrigerant pressure in evaporator to maintain desired temperature in cold room.

nection with coils of the direct expansion type regardless of the type of liquid refrigerant feed, is control by means of modulating the suction pressure in the coil in response to the room temperature. This is accomplished by the installation of a back pressure regulating valve in the main suction line from the coils in any particular room. The back pressure valve is actuated by a modulating electric or pneumatic type motor similar to one used for operating automatic dampers. The damper, or control, motor is modulated by a thermostat set at the desired room temperature. The suction pressure is thus modulated to maintain a constant room temperature. This method allows the complete use of all the coils in the room at the highest refrigerant temperature consistent with holding a constant room temperature. By the maintenance of the highest possible coil temperature, the highest possible humidity condition will normally be held in the room which will usually be desirable. By the addition of a pilot solenoid valve to the back pressure regulator, complete shut off of the suction may be obtained for defrosting. When this is done, a relief valve should always be used to prevent excess coil pressure.

Brine Temperature Controllers

In plants employing brine circulation for cooling, the brine temperatures are controlled to preset limits in the machine room at the chillers. Modulating back pressure regulator valves as described previously also work well on brine chillers to maintain constant brine temperatures. As the brine flows to the various room cooling coils, it must be controlled to hold proper temperatures in the cold storage rooms. Since it is not desirable to stop brine circulation for control, some type of by-pass control is usually desirable. A three-way valve is quite often used, temperature controlled, to permit brine to flow through the cooling unit, or by-pass around the cooling coil to maintain proper room temperature. In some instances where a small special temperature room is required utilizing a brine temperature that is desired higher than the plant brine temperature, a brine blending system may be used. This requires an auxiliary pump at the room being cooled. The pump recirculates a certain portion of the brine through the cooling coil and by the use of proper flow valves for admitting chilled brine; almost any brine temperature may be maintained. This method of brine temperature control is particularly useful for a small cooler or other special temperature use in an isolated location where only low temperature brine is available.

Temperature and Frost Control

Any rooms operating at temperatures below 35°F should have some means of defrost other than room air. These methods are discussed in another chapter. Adequate controls to operate the defrost cycle need to be provided so that this may be accomplished automatically. These controls are usually set for time clock operation so that defrost is accomplished at certain preset intervals. Timers are normally employed that both initiate the defrost at set times and program the defrost cycle and put the coil back into refrigeration. Various relays, contactors and other related devices are used in conjunction with the timer as required.

As plants increase in size, the control system becomes more complex since there are more functions and combinations of machines to control. Care should be exercised in the design of a control system that controls are not installed that merely control controls. In other words, always make sure that a control system design is as simple as is consistent with the functions desired; using adequate controls to insure complete safety and also complete operating controls, but avoiding the use of controls that serve no useful purpose and only complicate the system.

Control Panels

Control systems are sometimes placed in rather elaborate housings or enclosures with many indicating lights, etc. Although this can be somewhat costly, it serves a distinct advertising purpose for the plant in addition to the utilitarian function, and this publicity value should be given some weight in estimating the cost. The appearance of a well planned and neat machine room with a complete indicating and operating panel is impressive to visitors and prospective customers.

In an automated plant where the operating crew is minimum, a trouble indicating panel can have advantages. These panel alarm systems are obtainable commercially and are usually composed of a main panel broken into a number of small spaces, each space covered by an opaque covering on which is lettered the function it indicates. Each failure switch in the plant may, if desired, have a corresponding space on the alarm panel and by proper wiring, whenever a failure occurs, a light may be made to operate behind the proper space to indicate just where the failure is located. Thus an operator or maintenance man can tell at a glance at the panel just where any trouble may be. A further step may be taken in connecting the alarm panel to a call system where a failure

Courtesy of Alford Refrigerated Warehouses, Corpus Christi

Fig. 11.3. Automatic Control Panels and Switchgear for Automatic Cold Storage Plant

Note various indicator lights, gages and other features. Lower covers removed to show wiring connections.

occuring when the plant is unattended will send out an alarm to a watch service or plant personnel by telephone line. Usually only prime safety controls are connected to the outside alarm system. During times when the plant is occupied, the alarm system may be switched to an audible signal located within the plant.

Temperature indicating instruments are a good investment in any plant. These can be of the automatic chart variety that will indicate a multiplicty of temperatures throughout the plant at regular intervals. The temperature charts should be filed in some safe locations. They can be very valuable if a question arises as to storage temperatures in some location over a period of time.

Control Centers

Most modern automated plants concentrate all electrical control and starting equipment in a load center type enclosure, or enclosures, located in or immediately adjacent to the machine room. These panels, making up the electrical system, contain the various starters, relays, transformers and switches for the entire plant. With this type of distribution, most designers will elaborate the system to the extent of using indicating lights and labels at various points on the panel so that equipment operation is indicated by the indicating lights. Also alarm lights are often used to indicate a shut off of equipment for other than normal operating shut offs. In the most elaborate plants, a circuit or flow panel may be used which indicates on a panel, the schematic piping layout of the plant with indicating lights at various equipment and valve locations so that a complete picture of the plant operation is shown on the panel.

Some areas are subject to power flicker at times, particularly where a plant is served by a long transmission line. This flicker normally only lasts for a fraction of a second, but in plants using undervoltage release, an interruption of only a cycle or two can cause shut down. The undervoltage release is a desirable control feature but current flicker can cause nuisance shut downs. To combat this, sustaining relays can be used with a timing delay of less than 1 sec which will maintain circuit integrity over a flicker period. No bad results will occur since plant operation will not decay to a point of strain in pickup after $\frac{1}{2}$ sec or so as set on the relay. Any current outage over the time set on the relay will allow plant shut down.

Switching Controls

In large plants, it is quite often an advantage to be able to switch various refrigerant compressors from one suction pressure to another. This switching can be accomplished by push buttons, if desired, operating various solenoid refrigerant valves. It is usually not desirable to try and incorporate the switching as an automatic feature since such switching is only done at infrequent intervals. When a machine is switched, however, it should automatically lock into the control system controlling the particular suction into which it is locked.

Variations in Automatic Control Patterns

Variations of automatic control are endless depending on what the designer and the owner desire. A good design will function correctly and with minimum trouble over long periods of time, actually operating the plant at peak efficiency when the controls are properly set. As stated before, however, an automated plant is not an unattended plant, particularly in large installations. Such a plant will operate for periods without attention but there must always be one or more operating engineers employed as regular personnel of the plant who thoroughly understand the control system and its objectives and who can keep the controls in proper adjustment.

BIBLIOGRAPHY

AM. SOC. MECH. ENGRS. 1954. Automatic Control Terminology. ASME, New York.

ANON. 1959. Flow Measurement, Instruments and Apparatus. Fluid Meters: Their Theory and Application. ASME Power Test Codes.

ANON. 1965. Calibration of Liquid-in-Glass Thermometers. U.S. Dept. Comm. Natl. Bur. Std. NBS Monograph 90.

ASHLEY, C. M. 1953–1954. Controls. In Air Conditioning, Refrigerating Data Book, 8th Edition. ASRE, New York.

ASHRAE. 1965–1966. ASHRAE Guide and Data Book. Am. Soc. Heating, Refrig., Air-cond. Engrs., New York.

CONSIDINE, D. M. (Editor). 1964. Handbook of Applied Instrumentation. McGraw-Hill Book Co., New York.

GIBSON, J. E. 1958. Control System Components. McGraw-Hill Book Co., New York.

GIER, J. T., and DUNKLE, R. V. 1954. Using the heat flow meter to study heat transfer. Refrig. Eng. 62, 63.

HAINES, J. E. 1953. Automatic Control of Heating and Air Conditioning. McGraw-Hill Book Co., New York.

SAVANT, C. 1964. Control System Designs. McGraw-Hill Book Co., New York.

TRESSLER, D. K. et al. 1968. The Freezing Preservation of Foods, 4th Edition, Vol. 1. Avi Publishing Co., Westport, Conn.

TRUXAL, J. B. 1958. Control Engineers Handbook. McGraw-Hill Book Co., New York.

WOOLRICH, W. R. 1966. Handbook of Refrigerating Engineering, 4th Edition, Vol. 2. Avi Publishing Co., Westport, Conn.

CHAPTER 12

Defrost Methods and Systems for Cold and Freezer Storage Rooms

NEED OF DEFROSTING

Any cold storage plant with storage rooms below 35°F will require some type of defrost system. Whenever the refrigerant temperature is below the freezing point by any appreciable amount, ice and frost will form on the evaporator surface. As time progresses, this accumulation will tend to increase until the complete space between evaporator fins will be completely blocked to the passage of air and no further cooling will take place. Some positive method of removing ice and frost from the coil surface must be employed that will accomplish this in a reasonably short time so that room temperatures do not rise above a safe limit.

BASIC DEFROST SYSTEMS

There are five basic types of defrost systems used most frequently in the modern cold storage plant. These systems are: air defrost; water defrost; electric defrost; hot gas or hot brine defrost; and continuous defrost. All of the above systems are used in one form or another in plants today. There are advantages and reasons for using each of the various systems. In many instances, the principal reason for using a certain system is the familiarity of the designer with a particular method. All of the various types will work satisfactorily if properly applied.

Air Defrosting

The simplest system used in coolers is the air defrost system. This method relies on air at a temperature above freezing for defrosting. In its basic form, the refrigeration is shut off and the unit fans allowed to continue running. The room air, which must be above freezing, flows over the coil surface and brings heat to the coil which melts off the ice and frost and allows it to drip into the coil drain pan and drain out of the room. This type of defrost utilizes a rather low grade of heat so that defrosting is relatively slow compared to other systems. It is ineffective in rooms below 35°F since the ambient air temperature is only a few degrees above freezing and each cubic foot of air contains very little heat that can be used for melting ice. Since the system requires rather long defrost periods, coils must be oversize to accomplish the total daily load requirements in less than 24 hr by the length of the defrost periods. In rooms above 35°F, defrost can be accomplished by air without much difficulty. With the unit fans operating to bring heat to the coil surfaces, there is a problem of moisture pickup in the room air in rooms where low humidity is desired. As the ice begins to melt and water flows down over the coil surface to the drain pan, a certain amount will be evaporated into the air stream temporarily raising the relative humidity in the storage room.

Controls for air defrost are extremely simple. They consist of a solenoid valve or other means of shutting off refrigeration to the room for a specified time and then allowing the refrigeration to resume. A time clock to perform this function makes a complete automatic system of defrost. Brine fed units can also utilize this system of defrost by shutting off the flow of brine to the coil.

Forced Air Defrosting

A second type of air defrost that is sometimes employed in freezers or coolers is the forced outside air defrost system. In this type of defrost, the cooling coil and fan are contained within an insulated enclosure with two sets of doors or openings. In refrigeration, the coil is refrigerated and air is blown over the coil into the refrigerated space. To defrost, the refrigeration is stopped in the coil and the coil closed off from the cold room. At the same time, outside doors are opened into the coil compartment and warm outside air is blown across the coil to defrost it. In cold climates during winter operation, some provision can be made to heat the incoming air for defrost. In this type of defrost, there is very little moisture regain to the cold room since the cooling coil can be isolated from the cold room until it has completely drained.

Water Defrosting

Water defrosting of coils in freezers and coolers came into vogue a number of years ago when it was discovered that water could be introduced into a freezer room to flow over an iced coil and that the water would defrost the coil and flow back out of the room without freezing. The secret of this was to move the water in considerable volume and with good velocity so that it was not in the cold area long enough to freeze. Installations have operated successfully where water was pumped as far as 100 ft in an uninsulated line in a freezer to a coil, flowed over the coil to melt off the ice and drained back out of the room another 100 ft through another uninsulated line. Water lines are pitched to drain outside the cold room and all lines should drain to the outside, through proper traps, when the defrost water is shut off. Water defrost can be accomplished by means of sprays over the cooling coils or by perforated water pans allowing the water to flow over the coil in an even pattern. Water sources can be from city water or from tower water or any source of clean water available. Heavily treated water from towers should be avoided or any water containing chemicals that could contaminate stored products.

Water defrost can be made automatic by the use of solenoid water valves and refrigerant valves and the proper time clocks and control system. Piping water over a large plant can prove somewhat expensive as well as the disposal system either through recovery piping or by other means. For this reason, it is not often used in recent plants since more compact and less expensive systems have been developed. Blast freezers and some large coils do use water defrost in new installations. The water defrost provides a very fast and complete defrost and is easily adapted to blast freezers and large vertical floor units. Other specialized freezer applications also utilize water defrost, sometimes in connection with hot gas or other means. Water defrost is probably best operated manually rather than fully automatic. Safety devices for automatic control operate well but even an occasional failure can pile a large quantity of water on a floor in a hurry and in a freezer this soon becomes a mound of ice. With manual operation, or semiautomatic with an attendant present during defrost, very little trouble will be experienced. In many applications where a very fast defrost is required, water becomes very useful.

Electric Defrosting

Electric defrost systems have become quite popular over the years. In this system of defrost, refrigeration is cut off to the cooling coil and heat is applied by electric heaters. In some instances, the heaters are external to the coil and operate by convection, heating the air that surrounds the coil which, in turn, melts the ice. In other cases, electric heater rods are run through some of the tubes of a coil thus furnishing a more intimate contact between coil and heat for a faster defrost. Electric defrost is an excellent solution for isolated coils located long distances from the machine room and can be made to operate very satisfactorily.

The simplest type of electric defrost coil consists of heater elements placed immediately below the coil. For defrost, the refrigeration and unit fans are stopped and current applied to the heater elements. The elements heat the air surrounding them and the warmed air rises over the coil, defrosting it. Heater elements are placed in a variety of positions in and around the coil bank to try to achieve more even and rapid defrost. Upon completion of the defrost period, the current is shut off to the heater elements and refrigeration resumed.

In some electric defrost coils, a louvre or door is placed over the front of the unit. During refrigeration, the air pressure from the fan or some other device keeps it open for air circulation. During defrost when the fans are off, the door closes and thus confines the heat from the heater elements more closely to the coil and accomplishes defrost in a shorter time than with the open coil with no door or shutter.

Courtesy of Krack Corp.

Fig. 12.1. Electric Defrost Cooling Unit

During defrost, door is closed and air circulated over electric heater and coil to defrost coil.

A third type of electric defrost unit places the coil and fan unit within an insulated box with a motorized door or shutter operation. During refrigeration, the doors are open and the unit refrigerates in a conventional manner. The door or doors are closed during the defrost period; the heater elements are turned on and the fan allowed to continue running. This circulates air within the closed box over the heater elements and over the coil and accomplishes defrost in a minimum time. After defrost, the doors are opened and the unit returns to conventional cooling.

Electric defrost is quite widely used, particularly in the small and intermediate plants. A minimum of maintenance is required and no alterations are required to the refrigeration system. In large installations, with multiple refrigeration units, wiring and power consumption can be a factor and should be carefully studied before such a system is installed. Electric defrost is also used in addition to other defrost systems for selected rooms or systems in some large plants.

Hot Gas Defrosting

Hot gas defrost involves the use of discharge gas from the compressors, the discharge gas being diverted from the condenser to various cooling coils in the plant. The hot gas is condensed in the cooling coil. In condensing, heat is extracted from the gas and this heat melts off the ice and frost on the outside surface of the coil. Automatic valves are nor-

Courtesy of York Div., Borg-Warner Corp.

Fig. 12.2. Ceiling Type Unit Cooler Equipped for Hot Gas Defrost

Note hot gas coils welded to underside of drain pan to prevent freezeup of water in pan during defrost.

mally used to raise the pressure in the coils during defrost to maintain a condensing temperature above freezing so that ice will melt. A pressure relief valve should always be used to prevent excess pressure build up that would damage the coil. Various types of hot gas defrost are used in many plants from the smallest to the largest. It lends itself well to automatic operation. Since it is actually part of the refrigeration system, it responds to the same controls and safeties as used in the automatic control of the refrigeration system. It is a fast method of defrost since the heat is applied from the inside of the coil and exerts a uniform melting effect. In recent years, this type of defrost has come into favor, particularly in large plants using liquid refrigerant recirculation systems since the hot gas method works very well with this type of refrigeration system. In the small and intermediate plants, various patented systems are used where the single coil, single compressor type of installation may be defrosted.

In large multi-machine and multi-coil systems, hot gas is applied to certain coils and the liquid refrigerant condensed in these coils during defrost is distributed to other coils still under refrigeration by pump or other pressurizing means. In this type of operation, the disposal of the liquid refrigerant poses no problem if the coils undergoing defrost are kept to about 25% of the total tonnage of coils operating in the plant. This is a maximum figure and lesser tonnages of coils can be defrosted as desired. Defrost can be entirely automatic and programmed so that only selected coils are defrosted at one time. This type of system is particularly adaptable to larger ammonia plants.

Various schemes are employed in the small systems, ranging from a bleed system of condensed liquid refrigerant into the suction line, thus utilizing the heat of compression to dry the gas and supply defrost heat; to a system where heat is stored, from the discharge of gas, in a separate vessel in a secondary fluid and this fluid used to supply heat to evaporate the condensed liquid refrigerant during the defrost period. These two systems are used with halogen refrigerants. Other similar cycles are also in use. All seem to work well when properly installed and have an advantage of good operation under automatic control.

Brine System Defrosting

In brine systems using a brine solution as a refrigerant, the defrost systems employing air defrost, water defrost and electric defrost can be utilized in the same manner as for direct expansion. The substitute for hot gas is the use of hot brine. Part of the brine charge is run through a brine heater. Very often the heater is a heat exchanger mounted in the discharge line of the refrigeration system so that the hot discharge gas of the refrigeration system will heat up the brine. The coil to be defrosted is blocked off from the cold brine circulating system, either by the use of solenoid valves or by hand valves and the hot brine is allowed to flow through the coil through access valves, thus defrosting it. The cooled brine is returned to the brine heater and reheated. The hot brine forms a separate circuit with its own piping both to and from the various coils to be defrosted. The hot brine is normally circulated by means of a separate pump.

In a system employing continuous defrost, a low freezing point brine solution is continuously sprayed over the refrigeration coil surface to keep it clean of any ice accumulation. The wetted coil surface is also somewhat higher in heat transfer than a dry coil. The brine solution is quite often made up with common salt (sodium chloride) for cooler operation, particularly in meat coolers. It can also be made up of other solutions such as a glycol solution used in freezer operation. In the salt solution, as the cooling unit operates, moisture will normally be condensed out of the air and this will tend to dilute the sprayed solution. The brine solution is accumulated in a sump pan beneath the coil and to keep the strength of the solution high enough to prevent freezing, block salt is quite often placed in the sump and allowed to dissolve into the solution. Enough blocks are placed in the sump to keep pace with the rate of dilution. There is normally a continuous overflow from the sump pan due to the condensed moisture from the air.

A more sophisticated patented version of the brine spray unit is used in freezer operation. The pumped brine solution sprayed over the cooling coil is made up with a glycol base. As the brine is sprayed over the coil, moisture is condensed so that there is a continuous weakening of the brine, the same as in the brine spray unit described above. However, the weakened brine from the sump tank is circulated through a concentrator which separates water from the solution and returns a concentrated solution back to the spray unit. A source of heat, usually steam, is used in the concentrator to separate the water from the solution. The system can be made to operate automatically. Several cooling units may be operated from one concentrator. Initial expense is relatively high for this type system. It has the advantage of full time operation since it operates continuously with no time out for defrost. The system is used most effectively in areas where frost buildup is rapid and difficult to control.

Some method of delay is usually desirable, particularly in freezers, to bring on the fans after defrost. This delay should cover the time period to cool down the coil surface, at least to room temperature or below. If the fans are brought on immediately after defrost while the coil is still warm, a considerable quantity of warmed air will be blown out into the room. The expansion of this warmer air will tend to exert pressure on the walls and surfaces of the room, particularly the small and medium sized tightly constructed vault. This pressure can be sufficient to blow the vault door open in some cases or can cause cracking of plaster or similar finish walls. A simple thermostat at the coil outlet, strapped to the suction line or brine outlet, can hold the fans out of the operating circuit until the coil temperature has been reduced to the system operating temperature.

In an automatic plant, the selection of the type of defrost to be used is important and should be studied with care. In some instances two or more types may work out to advantage in the same plant. Some systems are patented and some are not and this feature should be checked. All types can be made automatic although some work out better under the design peculiar to a particular plant.

BIBLIOGRAPHY

TRESSLER, D. K. et al. 1968. The Freezing Preservation of Foods, 4th Edition, Vol. 1. Avi Publishing Co., Westport, Conn.

WOOLRICH, W. R. 1966. Handbook of Refrigerating Engineering, 4th Edition, Vol. 2. Avi Publishing Co., Westport, Conn.

CHAPTER 13

Operation and Maintenance of
Cold and Freezer Warehouses

OPERATION AND RECORD KEEPING

Operation of a warehouse, public or private, does not vary greatly except in record keeping and the dispensing of these records. In the case of the private warehouse, records are kept purely for the benefit of the individual owner whose merchandise is stored in the warehouse. This type of operation would be applicable to the food processor who owns the cold storage facility and stores only his own products both before and after processing.

PUBLIC WAREHOUSING RECORDS

In public warehousing, records are maintained by the warehouse owner, or operator, for the benefit of the customers who store merchandise. These records must be accurate and up-to-date at all times. Quite often a warehouse will be a combination of both private and public warehousing where an owner sets up space for his own operations and builds additional space over that required for his own use for public warehousing.

Variations in Records

In public warehousing, variations in service available are almost endless. The warehouse has the responsibility of keeping accurate records as to the whereabouts of the customers merchandise in storage. These records should contain complete information of the nature of the stored product; where it is stored and in what quantity. As merchandise is withdrawn, the records should so indicate and a balance made available indicating what part of the lot of merchandise, if any, still remains in storage. To be complete, the customer should have available to him at any time required, a complete inventory record showing merchandise on hand. In small and medium operations, these records can be kept by simple methods from hand prepared sheets to bookkeeping machines and not too much difficulty will be experienced. As the operation of a warehouse grows larger, more complicated systems are required and in the very large public warehouses, some type of computer system may be required. With this type of record keeping, almost any information relative to merchandise in storage is available on short notice. Inventory records are normally the base information fed to the computer. As additions and withdrawals are made to the merchandise in the warehouse, the record is stored in the computer system so that an up to date record is available at any time. Teletype machines in various work areas are often used in larger operations to transmit and receive orders and bills of lading for direct shipments out of the warehouse. Each warehouse operation is somewhat unique and record keeping must be tailored to fit the individual job rather than trying to use a perfectly standard form. No two warehouses will store various commodities in the same ratio. Services required will also be different with different localities. Because of these differences, a study of each operation must be made and a system developed peculiar to the needs of the particular warehouse being studied. Accuracy of record keeping is of prime importance, both from the standpoint of the customer and the owner.

CREWS NEEDED TO CONTROL FREEZER STORAGE OPERATION

Certain crews must be set up in a warehouse operation. The duties of each crew will differ depending on the area, labor agreements and other factors. Cars must be unloaded, merchandise palletized and stored in both dry and cold storage. As the merchandise is unloaded, it must be checked in and lot numbers assigned before storage and all of these records transmitted to the central record keeping area. The reverse sequence applies when merchandise is shipped out of the warehouse. Some warehouses specialize in carload movement where large quantities of merchandise are moved at one time while others specialize in smaller quantity movements.

TRANSIT STORAGE

Transit storage in which merchandise is stored in full carloads for an interim period and then shipped out to a final destination is a popular form of storage in some areas. Freight rates are adjusted to make this type of storage advantageous. Warehouses can also act as a distributing point for local shipments and transhipping in small lots from a large single lot storage. There are many variations in services rendered by various warehouses and each individual house seems to find its best service by experience and a continuous search for better ways to perform whatever services are required.

MECHANIZING THE REFRIGERATED WAREHOUSE

Almost all merchandise is handled mechanically in the modern warehouse and construction design is oriented to this type of operation. Fork-lift truck operation is standard in most instances with other supplementary types of handling equipment used along with the fork-lift trucks.

In the larger warehouses, loading and unloading crews are used both for rail car and truck operation. These crews may consist of from two men up depending on the methods of loading and unloading and the local labor situation. Fork trucks work with the crews to keep merchandise moving in or out as it is loaded or unloaded. Maintenance crews are also an essential part of the warehouse operation. In a large installation, there is constant breakage and wear of the items used in the operation and a good maintenance or repair crew is vital to smooth operation. Maintenance is essential in the machine room, on the materials handling equipment, outside rolling stock and on the building itself.

The Maintenance Crews in Cold and Freezer Storage

Maintenance crews are usually designated for certain jobs. This is normally dependent on the local labor situation and the organizational setup of the warehouse. Ideally, a crew should be able to pitch in wherever maintenance is needed regardless of the area or type of work required. Unfortunately this is rarely accomplished and crews are usually divided into various areas of work.

Machine Room Maintenance.—Machine room maintenance should be mostly preventive type of maintenance. The chief engineer or other designated individual should have a full set of records on all major pieces of equipment in the plant concerned with refrigeration. These records should contain a full description of the equipment with appropriate serial numbers. Spare parts lists should also be on file to be used in connection with the records of the equipment. This record should also contain oiling or greasing periods and other pertinent information about the equipment. If the equipment is of the type that needs to be inspected or overhauled at stated intervals, this information should also be contained in the records as well as space to indicate the date of the performed service. It should be the duty of the maintenance crew concerned with the machine room to perform the required preventive maintenance as required and to conduct routine overhauls of the machinery.

Overhaul procedures may include a minor or major overhaul. The time periods should be worked out and indicated in the records mentioned above. It is possible, that in some instances, too much maintenance in the form of major overhauls is performed on machinery, when only a minor overhaul and thorough inspection are all that is actually required. Preventive maintenance is important in any plant but care should be exercised that the periods of inspection and overhaul are kept reasonably spaced and not performed oftener than is required to assure good operation. Large heavy duty equipment will perform well over long periods of time with only minor inspection and attention. In smaller plants, it is sometimes less expensive to bring in an outside crew for overhauls rather than maintain an internal crew for this purpose. In larger plants, this is not normally true since there is always some routine performance check to be made by the maintenance crew.

A master chart, hung in the machine room, is desirable. This chart should list the oiling and greasing specifications for all of the plant machinery as well as the lubrication interval. Blank spaces should be included in the chart in which may be written the date that lubrication was actually performed for each item of equipment along with the initials of the person actually performing the lubrication. In a large plant, such a chart can be quite lengthy and may require multiple pages. The larger the plant, however, the more important such a guide becomes since the large plant usually has a large number of motors and rotating machines in operation and many of them scattered in remote sections of the plant. Without some method of keeping records of the lubrication periods, some motors or machines are bound to be neglected and even forgotten with a subsequent loss due to bearing failures from lack of proper lubrication. Charts on all machinery that must be inspected and overhauled or be lubricated at stated intervals are good insurance and also a valuable record for the plant chief engineer that maintenance or lubrication has been performed at the proper intervals.

OPERATION RECORDS

A set of operating records is almost mandatory in a large plant. Too many operators do not pay attention to temperatures and pressures set up by various items of equipment. In the refrigeration cycle, a set of temperature and pressure readings can tell a great deal about the operating efficiency of the plant. In the design of any plant, it is important that adequate thermometer wells, gauge connections and other adjuncts for taking readings be incorporated in the design so that a full set of pressure and temperature readings may be taken as desired. This data may be taken by direct reading of the various instruments or on recording instruments. At any rate, no matter how the data is taken, it should be reviewed constantly as a comparison of plant operation. Deviations from the normal readings should be investigated at once as these may indicate trouble. This review of the data is normally the job of the chief engineer and it is he who will investigate and direct whatever action should be taken.

OVERHAULS

Cold storage room units will operate over long periods with no major overhaul but they should be checked periodically and bearings should be lubricated at stated intervals. It is also important that the correct grade of lubricant be used in cold rooms, particularly freezers, since improper types can congeal and cause bearing failure. Other types of lubricants not suited for cold operation can lock bearings and cause motor troubles.

CHEMICAL TEST OF SECONDARY REFRIGERANTS

In plants employing brine systems, periodic checks of the brine pH value and density should be made and the necessary treatment added to keep a brine that is neutral and to prevent any corrosion. In addition to keeping a neutral brine, it is important that the density of the brine be kept at a reasonable point. The density should be kept to the minimum consistent with the lowest temperature expected in the brine system. As brines become more dense, particularly calcium chloride brine, the pumping horsepower becomes greater. Also the specific heat of the brine solution decreases so that more volume of brine must be pumped to perform the necessary refrigeration. This also will result in increased power required and higher operating cost. If the brine density is carried at a density of too high a point over the design conditions, plant temperatures become difficult, if not impossible, to maintain.

ROOM TEMPERATURE CONTROL

Room temperatures in the cold storage plant are important and a record of these temperatures at regular intervals is important. Customers pay for storage at certain temperatures and it is up to the warehouse to maintain these temperatures. Most modern cold storage plants do a good job of maintaining temperatures and the best proof of the temperatures held is a strip chart or other recorded data of temperatures in the various rooms. A good temperature recorder that automatically records temperatures in the various rooms at stated intervals is a good investment and a necessity in most plants. This type of temperature record keeping removes many doubts that might exist as well as inaccuracies that creep in with a manual-visual method of taking temperatures. In some specialized rooms, the relative humidity may also need to be recorded. This is usually done with an isolated recorder serving one area only. In the chart-type temperature recorder, machine room temperatures may also become part of the data taken.

The machine room usually operates with two crews; operating and maintenance. The operating personnel are on duty to see that the machinery operates properly. This crew may be operated in shifts over the entire 24-hr period in some plants either from necessity or where required by law. In automated plants, no operators are actually required, as a rule, except during the regular plant working hours. Completely automated plants will operate without trouble without attendance in off periods when the plant is not in business operation. The operators who are on duty during the regular working hours of the plant must be of a top flight variety with a complete understanding of the plant and its entire operation. The maintenance crew for the machine room should work in close cooperation with the operating personnel so that a close liaison is maintained at all times to keep the plant in top operating condition. The maintenance crew should be familiar with all of the machinery and also have electrical personnel who can service the control system as well as the plant electrical power and lighting systems. In smaller plants, many of the maintenance services required can be performed by outside service companies; but as plants grow larger, in-plant crews are usually favored.

GENERAL MAINTENANCE

The general warehouse maintenance crews are those who perform maintenance and repair on material handling equipment, warehouse equipment and building repairs. The size of the house will generally determine the size of the crew. No two warehouses will operate in exactly the same manner so that the number and personnel of crews will differ from one warehouse to the next. Fork-lift trucks require rather constant maintenance. Most warehouses with cold storage will operate battery powered fork trucks in cold rooms due to the hazards of exhaust gases from internal combustion engines in tight rooms.

Batteries require battery chargers and the chargers require electrical maintenance. There is a considerable amount of motorized equipment used in most warehouses so that a small crew is usually employed on the maintenance and repair of this type of equipment. Warehouse property itself is subject to abuse. Automatic doors are quite frequently hit and jammed. Walls can be damaged from fork trucks and other equipment and there is a constant requirement for maintenance and repair of building items. Some of the building crew can sometimes also be used in a carpenter shop for the repair and maintenance of pallets used in the cold and dry storage rooms. Pallets do get broken quite frequently and this item, in a large plant, can provide rather steady work for a small crew.

CLEANLINESS IN FOOD STORAGE WAREHOUSES

General housekeeping requires a few individuals and some mechanized equipment for sweeping and also, in some plants, trash hauling equipment. The maintenance of neat clean premises is an important feature in any plant. Large plants can accumulate a considerable amount of waste material that must be disposed of and it is no small task to keep a plant neat and clean.

There are other areas requiring maintenance such as painting, office maintenance, etc., and these areas are quite often handled by outside contracts rather than with plant personnel.

Maintenance and Repair Responsibility

All of the maintenance and repair work should be coordinated by one individual so that the work load may be spread evenly and equitably between various departments. This individual, quite frequently, is the chief engineer. It is his responsibility to see that the plant operates and functions as it should and to coordinate with the various sections of the plant to do maintenance work when it will be the least disturbing to the general operations of the plant. It should always be borne in mind that the plant is operated to the end of making a profit. This profit can only be achieved by having satisfied customers. Maintenance is important to keeping the customer satisfied and is a good tool in the overall operation of the plant.

CHAPTER 14

Record Maintenance in Cold and Freezer Warehouses

INTRODUCTION

The cold and freezer refrigeration warehouse as a privately owned, public agency, has obligations much beyond the quarterly reporting to its board of directors. From the organization standpoint the management must keep the Internal Revenue Service informed at least annually of all of its income-outgo. Since they are handling food in large quantities, usually on the contract basis, both in transit and for local delivery, they must keep adequate records for the Food and Drug Administration, for the U.S. Department of Agriculture, and Commerce, and Interior.

Within the larger family of cold and storage warehouses of the United States they should maintain close contacts with the National Association of Refrigerated Warehouses, Inc., Washington, D.C. This association maintains general contacts with its member warehouse managers through the "House" monthly magazine, "Cold Facts," and the publication of suggested rate structures reviewed in their "Rate Manual" and Storage Package Rate Tables.

THE RECORD SYSTEM

There are many systems of record maintenance and reporting available on the market for American managers of cold storage warehouse enterprises. Unless the president or executive manager of any industry is himself a certified accountant or an auditor of experience, he should not attempt to design a system that will meet the approval of the taxing and banking agencies with whom the warehouse must cooperate for his operations and his financing.

Most acceptable auditing firms have their own forms, system and auditing sheets, that they have found most useful in serving in their capacity as certified public accountants and unless the business has an already going established record system that has been found most adequate, any new or replacement system should be introduced with the full cooperation of the accountants concerned.

Most helpful recommendations and suggested systems for refrigerated warehouses accounting can be obtained from the National Association of Refrigerated Warehouses.

RECORD MAINTENANCE

In the supervision, maintenance and operation of any business, record keeping is not only a requirement now imposed upon all industry and businessmen by the Internal Revenue Service, but has been for several generations one of the principles specified for the successful, efficient, operation of any industry that expects to compete successfully.

While cold and freezer storage warehousing may or may not have within their daily activity, manufacturing of products as such, there is an obligation to maintain a very exact income and outgo of all products, labor, electrical energy, office keeping and the obligations they have accepted in the storage liabilities within their business enterprise. These records should be kept from the first day they take over the ownership and supervision of the cold and freezer storage business. Some of the more critical items of which the records should be maintained, include: (a) the cost of the money required in establishing the business; (b) meeting the daily payrolls; (c) cost of equipment and materials to keep the warehouse operating efficiently; (d) the outlay for maintenance; (e) continuing repairs; (f) power, light and air conditioning; (g) the cost of salesmen and general personnel and department management; (h) city, state and county, and Internal Revenue taxes available when demanded by the several agencies; (i) the cost of accident, fire, tornado and theft insurance, a continuous cost of owning any industry.

The every day cost may be divided into two main categories: (1) owning; and (2) maintenance and operation. The owning will include amortization, interest and insurance.

A breakdown of the operating expenses will include all energy demands, all maintenance, operating personnel, management personnel and all utilities service. It will not include those utilities not classified as power.

Service and Energy Costs

In the annual audit very exact accounting of all power and lighting is essential; these costs are very often related to the material cost of water, gas and oil, depending upon the source of power utilized by the management.

Air, Water and Sewerage Control

This item is becoming a very important issue in every one of our modern cities. Several cold and freezer storage facilities of our American cities have moved out of the city limits of our best regulated municipalities. This was to avoid meeting the requirements imposed by those cities for adequate control of air and water pollution. Every manager of a cold and freezer warehouse, especially those permitting food processing on their premises, should be certain that his plant design includes a satisfactory method of sewerage clarification that will meet municipal, state health and national requirements as prescribed by law.

Operational and Management Salary Schedule

In the successful management of cold and freezer storage enterprise there is no place for "window dressing" salaried employees. The budget of a well operated enterprise in the cold and freezer storage industry to be competitive must be organized anticipating that from the manager or owner to those employees assigned to cold rooms, docks, and to the maintenance of records and similar clerical assignments, these are all performed by active "working" personnel.

The items that must be recognized in the monthly and annual work sheets for the Management and Operation of Cold and Freezer Storage Warehouses are given herewith in Table 14.1.

MANAGEMENT AND OPERATING COSTS RECORDS

Many business failures in the operation of cold and freezer warehouse enterprises can be laid upon the failures to anticipate all of the operating and management costs that accrue from the amortization of the original investment and to the annual monthly repetitive costs of doing business on a scale acceptable to the public.

Auditing Control and Amortization

The manager and/or owner of any cold and freezer storage warehouse must maintain comprehensive cost accounting procedures. They should be set up so that the owners may recover each year a portion of the initial cost as an expense item charged against revenue. This decrease in the value of property should be determined by the company auditor in conference with the Internal Revenue observers and all US Government control agencies related thereto officially delegated to control the quality of food, disposal of rejected foods, and sewerage

TABLE 14.1

ITEMS THAT MUST BE RECOGNIZED IN THE MONTHLY AND ANNUAL WORKSHEETS
FOR THE MANAGEMENT AND OPERATION OF THE COLD AND FREEZER
STORAGE WAREHOUSE

(1) Investment Amortization (after annual)

cost of storage facilities
cost of the mechanical and electrical system
cost of the management facilities

(2) Annual Overhead

interest
income taxes (IRS)
property taxes
insurance
rental charges

(3) Maintenance Costs

replacements
water and oil
painting
refrigeration
salaries of maintenance men

(4) Service and Energy Costs

power for pumps
power for refrigeration machines
power for fans
lighting and miscellaneous motors

(5) Materials Costs

water
gas
fuel oil
other oils

(6) Air, Water and Sewerage Control and Clarification

air filtering and cleaning
water treatment
sewerage disposal and clarification

(7) The Operational and Management Salaries

and the financial structure of the corporate setup. The auditor and the management should have full agreement with the Internal Revenue observers, banking interest, and Federal agencies assigned to protect the public interest in the affairs and procedures of the food dispensation and storage. In all of this the auditor and management should give careful consideration to the amortization of the plant and facilities.

Some items that must be included in amortization that may determine the length of period agreeable to all, are: (a) type of building and its useful life; (b) type of equipment, and (c) the lease or ownership conditions.

Annual Overhead.—This item is primarily the interest that must be paid annually and the tax items that must be paid both to IRS and the property taxes due the city, county and state, plus the insurance and rental obligations annually recurring.

Maintenance Costs.—The manager of any cold and freezer storage plant should recognize that he is in a commercial enterprise that is highly competitive and the maintenance of his plant can be one of his best advertisements.

TABLE 14.2

RECOMMENDED DEPRECIATION PERIODS[1]

Item	Years
Air conditioning systems	20
large—over 20 tons	20
medium—5 to 15 tons	15
small—under 5 tons	10
Air washer (see Dehumidifier)	
Compressors	
air	20
Condensers	
double pipe	20
evaporative	15
shell and tube	20
Coolers, water-tank and coil or shell and tube	20
Dehumidifier	10
Drums, purge or surge	20
Ducts and other sheet metal work[1]	± life of building
Engines (gas, diesel or dual fuel)[2]	20
Fans	15
Filters, air, oil, self-cleaning	20
Gages	15
Heaters	
gas	15
water heaters, open or closed type	20
Heating systems	
boilers and furnaces	20
Insulation	15
Motors	
induction indoor	20
induction, weatherproof for outdoor	20
Synchronous and exciter set	20
Piping, refrigerant and other	20
Pumps	20
Receivers, refrigerant	25
Spray pond	15
Switchboards, electric	25
Thermometers, room type or recording	15
Tower, cooling	15
Transformers	25
Turbines[2]	20
Valves	
relief	20
automatic expansion and by-pass	5
water regulating	20
Wells and well pumps	25

[1] From system evaluations studies by W. R. Woolrich and E. R. Hallowell.
[2] Estimated values, not IRS values.

BIBLIOGRAPHY

ALLEGRI, T. H. 1956. Materials handling of the frozen foods in storage. U.S. Dept. Agr., Agr. Marketing Serv., Marketing Activities *19*, No. 1, 11–14.

ASHRAE. 1967. Guide and Data Book. ASHRAE, New York.

BULLINGER, C. E. 1958. Engineering Economy, 3rd Edition. McGraw-Hill Book Co., New York.

GRANT, E. L., and IRESON, W. G. 1960. Principles of Engineering Economy, 4th Edition. Ronald Press, New York.

MINER, B. D. 1966. The basic principles of good management. National Convention of Locker and Freezer Provisioners, St. Louis, Mo.

WOOLRICH, W. R. 1968. The warehousing of frozen foods. *In* The Freezing Preservation of Foods, 4th Edition, Vol. 1, D. K. Tressler *et al.* (Editors). Avi Publishing Co., Westport, Conn.

CHAPTER 15

Lighting and Electrical Facilities for Cold and Freezer Warehouses

DEFINITIONS FOR LIGHT AND LIGHTING

Brightness

Brightness of any object is dependent upon the intensity of the illumination on it and the percentage of light reflected or transmitted. Surface brightness is the degree of luminous intensity of any part of the surface when viewed from any angle.

Fluorescence.—Fluorescent lamps are electric-discharge units in which the radiant energy of an electric discharge is transferred by suitable materials into wave lengths of higher luminosity.

Filament or Incandescent Light.—Filament lamps are light sources that consist of glass bulbs with electrically maintained filaments at incandescence temperatures from heat energy created from the electric current flowing therein.

Foot Candle.—One foot candle is a measure of illumination on a surface which is 1 ft distant from and perpendicular to a source of illumination of one candle power. The foot candle and the lumen are interrelated in that one lumen is the unit of light flux representing a quantity of light falling upon 1 sq ft of surface every part of which is 1 ft distant from a point source of one candle power. The scientific foot candle or "candle foot" is defined as "the illumination produced by a British standard candle at a distance of 1 ft."

Glare.—Glare has been defined as "brightness in the wrong place." More briefly, glare is dazzling brightness.

Lighting engineering terms it radiant energy emanating according to its capacity to produce visual sensation.

Mazda is a trade name for certain incandescent bulbs used by manufacturers.

Watt rating is a unit of electrical power for classifying lamps by consumption. A 40-watt lamp will require 40 watt hours per hour if turned on continuously.

Light and Vision

Light excites the sensation of sight and must fall on objects if they are to be seen. Shadows are formed by objects that cut off the rays of light creating the shadow. Illumination from a light source decreases with distance according to the law of inverse squares; that is, at twice the distance the source of illumination is $\frac{1}{4}$ times, and at 4 times the distance it is $\frac{1}{16}$.

Sunlight and Natural Illumination.—On top of a mountain at summer midday, with no clouds present, the illumination approaches 10,000 ft-c. Under a well shaded area, the illumination would be from 350 to 500 ft-c on the ground level during a summer period.

CURRENT AVAILABILITY OF LIGHTING EQUIPMENT

There is no effective standardization of lighting equipment. The three types found on the market are: (1) the incandescent filament; (2) the fluorescent; and (3) the mercury vapor. All three types are used both outside and inside with proper protecting and decorating fixtures.

Many needs for lighting around a cold storage plant such as for stairways, loading platforms and entry ways can be well illuminated with the simplest and less expensive equipment.

Within cold storage warehouses the lighting is much different from most industrial applications in that no sunlight is wanted within the cold storage walls, thus the best and most effective in-plant lighting with minimum shadows and heating yet maximum illumination are most acceptable.

The closest approach to natural light is obtained with fluorescent types of equipment and facilities within the storage rooms. In the halls and stairways, incandescent lamps are usually more functional. The office may find that a mixture of fluorescent and filament lighting will give the best distribution.

The heat given off by fluorescent light will be about 30 to 40% of that from the filament type lamps of the same candle power. Likewise the power bill per fluorescent candle power will be about 40% of that for the filament type of installations.

The life of fluorescent lamps depends mostly on the number of times the lamp is switched on and off rather than the burning hours. Fluorescent lights that burn continuously last 5 to 15 times as long as ordinary Mazda lamps. Even when turned on and off frequently, the fluorescent lamp will last from 3 to 5 times as long as ordinary lamps. The service brightness of fluorescent lamps provides diffused light approaching that of the shade under a tree at midday.

Fluorescent lamps can be obtained, made to order, in many lengths on the regular market, and they are available in 1.5 to 8 ft lengths, usually listed as 20-, 40- and 60-watt units. On the American market, the 48 in. 40-watt fluorescent is most popular.

The operating characteristics and make up of the fluorescent lamp unit includes the lamp, the ballast and the starter. The usual fluorescent lamp has a filament at each end which serves as a starter for a second or two when the light is first turned on; the ballast has an induction coil with a steel core designed to limit the total power and momentarily acts as a step up transformer.

Factors Influencing Seeing

Ability to see depends upon four basic variables: size, contrast, brightness and time of exposure. It is understood, in considering the factor of size (a function of the angle subtended by the eye), that either the distance is assumed to be fixed or the size is assumed to be adjusted if the distance is varied. It would be justifiable to include distance as a factor, but it is less complicated to consider the foregoing relationship between size and distance.

Size, for seeing measurements, depends upon the visual angle (normal vision, 1 min; best vision, 40 sec), and visual acuity is measured by the minimum size seen. The absolute size cannot ordinarily be altered, but the visual size may be altered by changing the distance.

Contrast is a relative matter and may be best expressed in percentage by

$$\text{percentage of contrast} = \frac{\text{background brightness} - \text{object brightness}}{\text{background brightness}}$$

Expressed in this way, the contrast of a perfectly nonreflecting body superimposed on a body reflecting all light will be 100%.

Brightness indicates the average brightness of the object. If the surface is mainly background, the brightness approaches the brightness of the background; but, if the surface is composed of a combination of objects or bodies of various brightnesses, then brightness means the apparent brightness.

Time of exposure is a factor of great importance, for the longer an object is exposed to the retina, the more clearly can its presence and its details be distinguished. This is particularly true in most production tasks and in automobile driving.

The influence of these factors upon the ability either to see or to recognize an object is self-evident. The inadequacy of any one item must be compensated for by the others if the same ease of seeing is to be maintained. By adjustment of the four factors, unless the object is microscopic, the object may be made visible to such a degree that the task may be performed upon it with the least effect upon the nervous system of the subject. In most instances all but the brightness is fixed; therefore, the importance of the problem of adequate and proper illumination is of prime consideration.

Probable Factors Influencing the Effectiveness of Lighting Systems

The development of systems for artificial lighting is continually changing. The recommended amount of illumination, the methods of control, and the costs have necessarily changed together. Today electric energy costs less, illuminants cost less, and the efficiency of the illuminants is higher; therefore, it is possible to recommend higher levels of illumination in order to make the task easier, without increasing the cost.

In summary, it may be said that those factors contributing to a sucessful lighting system are:

(a) Quantity of illumination for proper brightness
(b) A good quality of illumination
    (1) no direct glare from the source
    (2) a minimum of reflected glare from the work or surroundings
    (3) uniform lighting on the work surface
    (4) illumination free from shadow
    (5) surroundings illuminated with the proper brightness contrast
(c) Comfort
    (1) elimination of opinions based on past experience
    (2) sufficient time to determine the reaction to the illumination supplied
(d) Work surface considered with relation to lighting

The Quantity of Illumination

The eye, which acts as a camera does, has the same essential parts: a cornea, an iris, and a retina, corresponding to the lens, the shutter, and the film of the camera. Each part of the eye has a special function and has muscles to control the adjustments. The lens is shaped by the ciliary muscles and focuses the image upon the retina; the iris is closed or opened to admit more or less light, and the muscles attached to the eyeball focus the eyes upon the same object. These functions are classified as: (a) accommodation—lens adjustment; (b) adaptation—iris adjustment; and (c) convergence—focusing both eyes upon one object.

The retina is the light-receiving mechanism, and its nerve system, in conjunction with the brain, reproduces the visual image focused upon it.

The eye has evolved from a rudimentary mechanism until it has a sensitivity far beyond that of any device man has ever created. On bright summer days the illumination in the open may reach 10,000 ft-c, which is, at best, not too comfortable, but 1,000 ft-c, under the shade of a tree is not objectionable. At the other extreme, a bright, moonlight night has an illumination of approximately 0.04 ft-c, and some people make claims that they can read a newspaper with this amount of light.

The eye has the ability to adjust itself to a wide range of illuminations, and this particular ability is the worst enemy of those wishing to correct faulty illumination, for it is a common belief that the illumination is satisfactory if it is possible to see with any degree of comfort. The last hundred years have brought the worker from the outdoors into buildings and from tasks utilizing distance vision to tasks most of which are performed about 14 in. from the eyes. All these changes have been accompanied by a decrease of illumination—almost 90% in most instances.

Brightness

(1) When the light output of a source is increased 10 times, its brightness should be decreased to not more than $\frac{1}{2}$, or perhaps even to $\frac{1}{5}$, of its former value; in other words, the light output of a source cannot be increased as fast as its area, but it is safe to increase the output as fast as the diameter.

(2) Increasing the general level of lighting in a room ten times permits only doubling the brightness of the light source.

(3) Doubling the mounting height of a unit permits increasing approximately 3 times the lumen output of the lamp which may be used in the unit.

Glare

When brightness becomes an annoyance, it is called glare. Glare is any brightness within the field of vision causing discomfort, interference with vision, or eye fatigue. It is a sensation governed by the surroundings and is variable with the individual; therefore, it is subjective. Much research has been directed toward a study of the effect of glare, but glare is very difficult to measure. Glare may be caused by: (a) high brightness of the source; (b) high contrast between source and background; (c) location of source in field of view; (d) total volume of light entering the eye; and (e) time of exposure to the source.

Correction may be made by removing the offending source from the line of vision. The angle above the line of vision at which the source should be mounted is approximately 14° to 18°. As stated before, the visual size of the source is proportional to the distance from the observer.

Glare may be classified as direct and reflected. Usually the annoyance from direct glare is so marked that immediate steps are taken to correct the cause; reflected glare, however, is much more subtle and may do damage without awareness of its presence by the victim until the physiological effects become acute. Reflected glare is sometimes called glint, and this property is used to recognize small objects and to inspect material; in this instance the specular reflection of the light is utilized.

Foot-Candle Prescriptions

Though illumination is measured in foot-candles in an objective manner, the prescribing of the amount to use is strictly governed by subjective analysis. The amount of illumination needed is only one factor, the other being economy. Economy has not been a limiting factor, for, considering the cost of lamps, the increase of lamp efficiency, and the reduction in power rates, it is possible to obtain today, for a given amount of money, 2.5 times as much light as could be obtained in 1925.

100 ft-c or more.—Obtained by supplementary lighting where the task is severe and prolonged, the details are very small, the contrast is low, and the speed of operation is high.

50 to 100 ft-c.—Produced by supplementary lighting for close work where speed is not a factor and the work is small in size.

20 to 50 ft-c.—Preferably supplied by local lighting if convenient, for the range is the upper limit of general illumination. The normal industrial and commercial tasks for close desk and office work fall into this group.

10 to 20 ft-c.—Applicable to most recreational needs and ordinary tasks that are not prolonged throughout the working day; easily obtained with general lighting.

OUTSIDE LIGHTING OF DOCKS AND YARDS

Outside or exterior lighting may include yard, roadway and dock illumination, advertising installations and police protection lighting of the corporate properties. Since many deliveries arrive at night or early morning, it is recommended that an auxiliary power source be available so as to not leave the property unprotected for regular collections and deliveries any hour of the day or night. Frozen food deliveries should be transferred into the freezer storage warehouse with as little delay as possible from the delivery trucks. For property protection and for effective frozen food preservation, yard and dock lighting on a "round the clock" schedule is recommended.

TABLE 15.1

RECOMMENDED ILLUMINATION IN COLD AND FREEZER STORAGE WAREHOUSES

	Foot Candles
Cold and freezer storage rooms	50–75
Hallways and stairways	10–20
Machine rooms	60–75
Auditing, tabulating and bookkeeping rooms	50–60
Drafting and designing rooms	60
Reception and stairways	20
Conference and active file rooms	30
Wash and locker rooms	30
Loading and unloading docks	60
Servicing yard and driveways	30

RATES FOR ELECTRIC SERVICE

The cost of electricity is based on different factors in different cities. These costs include the capital invested, a factor of readiness to serve; the time of the day or month the power is used, and the relation between demand and the amount actually consumed within a given period; these are factors entering into the rates prescribed in the contract. A rate may be lowered by off-peak loading and this is especially applicable to refrigeration loading.

Load limiting controls of several types are available and are involved in the rate making contract with the power supplier. With any type of control, time delay relays are installed in order that upon service restoration after interruption the individual circuit unit will not come on simultaneously but take up over a period of a few minutes.

ELECTRIC LIGHT AND POWER WIRING

The National Electrical Code Rules which are usually recommended by the National Fire Protective Association should be followed in any industrial and office wiring. In addition to the National Code, many cities adopt local codes that take precedence over the National Code. Most local electrical contractors are familiar with the recommendations within the area in which they do business. If an outside electric contractor is bidding on local work, he should become well informed of the recommendations legally prevailing in that area.

In general, the type of wiring to be used for the electric lighting and similar electric service will depend upon the location governing the installation. Within buildings, rubber covered wire and nationally approved plastic covered wiring may be used for so-called open wiring, and should be used in concealed work. Outside weatherproof wire is used with special types recommended for high heat and high moisture locations. Lead sheathed conductors are usually recommended for underground work.

Wire Sizing

The smallest size copper wire permissible for electric light and fire protection is No. 14 B and S wire, although No. 16 and No. 18 B and S wire is commonly found in the cheaper fixtures in households.

The maximum load permissible by the National Electrical Code, on a 110-115-v system is 15 amp and a maximum load of 1,725 watts. Usually the number of outlets on any 1 circuit is legally fixed at 10, although this may vary under some local conditions.

Ground Wires and Grounding

There is often confusion as between ground wire and grounding wire. This has led also to further confusion and controversy on bare neutral wiring.

In most regions grounding wire refers to a wire which does not carry current during normal operation. It is a grounding wire for security and is often tied to the water supply lines or other plumbing facilities to assure an adequate grounding of the system.

The grounded wire refers usually to the neutral wire of the circuit. Under some codes this neutral wire may be incased in the cable as a bare wire, thus the name bare neutral. Some local codes do not accept bare neutral wire in their area.

The National Electrical Code is sometimes supplemented by local codes or ordinances which are never contrary to the National Code, but limit its application. For example, armored cable wiring is one method permitted by the National Code, but sometimes prohibited by local code.

Permits

In many places it is necessary to get a permit from city, county, or state authorities before a wiring job can be started. The fees charged for permits generally are used to pay the expenses of electrical inspectors, whose work leads to safe, properly installed jobs. Power suppliers usually will not furnish power until an inspection certificate has been turned in.

Lighting Circuits

The Code requires enough lighting circuits so that three watts of power will be available for every foot of floor space in the house. Since a circuit wired with No. 14 wire and protected by a 15-amp overcurrent protection provides 15 × 115, or 1,725 watts, each circuit obviously is enough for 1,725 divided by 3, or 575 sq ft.

TABLE 15.2

CURRENT REQUIREMENTS FOR VARIOUS SINGLE-PHASE
MOTORS AT 110 AND 220 VOLTS

Motor	110 V	220 V	Motor	110 V	220 V
1/4 hp	6 amp	3 amp	1.5 hp	20 amp	10 amp
1/3 hp	7 amp	3.5 amp	2 hp	24 amp	12 amp
1/2 hp	10 amp	5 amp	3 hp	34 amp	17 amp
3/4 hp	14 amp	7 amp	5 hp	56 amp	28 amp
1 hp	16 amp	8 amp			

THE ELECTRIC GENERATOR

What was in the early decades of electrical history referred to as a dynamo, today is generally called the electric generator.

Virtually all electrical current, when generated, is alternating in form. The typical alternating current generator consists of a series of cutting loops connected to individual collecting rings. The collecting rings in turn are contacted by brushes from which the current is led to an external electrical circuit for distribution. To increase voltage picked up at the rings, the number of turns in each loop to form a coil connected to the ring, is increased. These coils are wound around an iron laminated core to form the circuit. The function of the iron core is to increase the magnetic lines being cut by revolving coils. The voltage generated in each coil is the sum of the voltages of each individual turn in the completed loop. The rotating armature is a mechanical assembly made up of a number of coils.

If this alternating produced current is needed flowing only in one direction then a commutator, made up of a segmented drum, each segment insulated against contact with each other, must be built into the machine. Carbon brushes will be installed on this commutator in a position positive and negative, respectively. Current will now flow through the wires connected to these brushes in one direction as direct current.

Depending, therefore, whether we use collector rings or a commutator in leading off the current produced in the generator, we will have alternating or direct current, respectively.

Alternating current is one in which the flow continually reverses itself, to travel first in one direction and then in the opposite. The principal phenomenon of alternating current, that is different from direct current, is that alternating current has the characteristics of inductance and capacitance.

Capacitance

In alternating current applications a condenser, or called by some a capacitor, is often found in the circuits. The typical condenser is made up of two or more plates of a conducting material separated by some form of electrical insulating plates. These plates may be made of paper, glass, porcelain, or some of the insulating plastics. Even air can be used as an electrical insulator between conducting plates of a condenser. A typical electrical condenser will block the flow of direct current but permit alternating current to go through. When the circuit is first closed direct current flow will proceed to each side of the condenser, then stop. In the case of an alternating current instantaneous flow will occur each time the current reverses direction and thus gives continuous energy flow.

The quantity of electricity in the condenser, momentarily stored up, will depend on: (1) the impressed voltage; (2) the size of the conductors; (3) the thickness of the insulator; and (4) the length or dielectric strength of the insulating material.

The ability of the electric insulator to prevent leakage is known as its dielectric strength. This strength depends on its thickness and the number of volts it will withstand per unit of thickness.

The characteristic of the condenser is to permit alternating current to pass through the condenser circuit but block off continuous flow; direct current makes it possible to separate out direct current when flowing in the same conductor as alternating. On the other hand, since direct current passes through an induction coil with only resistance it is quite possible to separate direct current from the alternating current when at similar voltages they all pass through a common conductor by "splitting" or branching the line and passing alternating current on to the condenser route and the direct current on to the induction coil route to their separate points of origin.

Inductance has been defined as the property of the electric circuit by which the varying current in it induces voltages in that circuit or in an adjacent one.

Conductance

That property of an electrical circuit which determines for a given applied voltage across its terminals; the average rate at which electrical energy is converted into heat. The unit is mho the reciprocal of ohm.

Impedance

In AC circuits, the factor of impedance must be considered in circuit computations. It is usually designated by Z. In direct current circuits there is little need to consider the Z factor. In the typical alternating current, impedance is the result of having either induction or capacity reactance in additon to the resistance of the circuits in ohms. It is a form of resistance occasioned by the induction and capacitance of a circuit.

TRANSFORMERS

The transformer is an electrical machine, usually without any moving parts. It transforms alternating voltage from a higher to a lower, and from a lower to a higher voltage. Usually it is made up of a special soft iron core and two electrical circuits called secondary and primary. The fundamental equation of the transformer is

$$\frac{\text{current in the primary}}{\text{current in the secondary}} = \frac{\text{number of turns of wire in the secondary}}{\text{number of turns of wire in the primary}}$$

Transformer Connections

Single-phase transformers may be connected as: single-phase, 2-phase and 3-phase. Usually if the system is 3-wire, 230 v circuit, the 3-phase circuit is grounded at the common connection to each phase.

Transformer Ratings

The rating of the transformer depends upon the heat generated in the windings and the associated iron core. In designing the transformer, this heating must be balanced against the ability of the machine to give off its heat as fast as generated. To offset this heat, most large transformers have their windings immersed in a special transformer oil. Provision then is made to keep this oil cool by either a water or an air circulating system.

Not always can oil be used for cooling on account of the fire hazard, in such cases if air cooling does not suffice, some noninflammable liquid is substituted for the oil.

ELECTRIC MOTOR DESIGN

The electric motor is designed with the two parts mentioned previously: the stationary poles or stator, and the rotating loop or loops called the rotor, these conveniently mounted in a metallic frame, designed to withstand the electric and mechanical forces inherent to the job.

Classification of Motors

The National Electrical Manufacturers Association, supported by the manufacturers of electrical apparatus, classifies motors by: size, applications, electrical types, method of cooling and the mechanical protection. Some manufacturers even classify their electric motors by the design letter or design number.

In general, electric motors have been rated on the basis of operating without heating at 104°F. More recent ratings of many motor supplies have been upward from the 104°F ceiling.

Another classification often used is by electrical type; thus, for direct current motors, they are usually classified: shunt-wound, compound-wound or series-wound. Alternating current motors have several classifications; squirrel-cage induction, wound-rotor induction, synchronous, series, and single-phase. Single-phase alternating current motors of the squirrel-cage type may be classified as: split-phase, resistance-start, capacitor-start and permanent split-capacitor. They may also be classified as repulsion or repulsion induction.

Three types of single-phase motors that the maintenance and repair men, in cold and freezer storage warehousing, meet most often are: the split-phase, capacitor, and the heavy-duty capacitor motors. The split-phase motor is commonly used for easy starting loads, where the maximum load is applied after the motor has attained full speed. It is extensively used on operating light power tools, paint sprayers and where the normal load is less than $\frac{1}{2}$ hp. They should not be used on a hard-starting machine such as a compressor or a pump.

Capacitor Motors

While the capacitor motor is similar in construction to the split-phase motors, it is equipped with a condenser which aids in starting of the heavier loads and reduces the starting current. Capacitor motors are suitable for compressor loads and for production equipment that must start under load.

Heavy Duty Motors.—Heavy duty capacitor motors are similar to the capacitor motor but have superior starting ability, especially for refrigerant compressors, deep well pumps and air compressors.

Polyphase induction motors are usually classified as squirrel-cage and wound-rotor.

Motor Rating

Motors are generally rated in horsepower. It is possible for a man doing heavy labor, to maintain for some time the equivalent of $\frac{1}{10}$ hp. A 1 hp motor is capable of exerting as much energy as 10 hard working men.

Electric motors are designed to deliver more horsepower when starting than running under normal load. Some motors are designed to deliver a heavier starting load than others. On overloads some motors are designed to carry on for some time with a 50% overload. Their temperature rise under load is a good index of overloading. It is of some advantage to have electric motors that can carry an overload for a short period, but they should not be pressed into regular service at such loadings.

Motor Speeds

Motors that operate continually at 900 rpm are usually much heavier than those that run at 1,750 rpm. The 1,750 rpm motor will be much cheaper than the 900 rpm units since they can be much lighter in weight.

Since the higher speeds involve much more inherent vibration, these motors must be machined to much closer micrometer limits. This highly accurate machining increases the completed motor price of the high speed motor somewhat but not in proportion to the saving in material when going from 900 to 1,750 rpm.

Several companies market 3,450 rpm motors for price competition, even up to 100 hp and larger, in the United States and Canada. European manufacturers seldom exceed 1,750 rpm except for some motors of less than 1 hp. There are very few American electric motor manufacturers that have the facilities and the expert machinists capable of producing 3,450 rpm motors of high quality in the United States, and their purchase should be avoided if the management of cold and freezer warehouses anticipate using such units more than 3 yr. Some American purchasers of 3,450 rpm motors of these larger sizes have experienced more than a 50% breakdown in less than 1 yr.

The Direct Current Motor

All electric generators of the dynamo type consist of two essential parts; the field magnets and the armature on which copper conductors are mounted so that they revolve between the field magnetic poles and thus cut the magnetic lines of force.

The simplest dynamo generator would be one single loop of wire rotating between the opposite poles of the magnets.

If these loops move at right angles and revolve through $\frac{1}{2}$ turn, or 180°, each side will cut the whole number of magnetic lines coming from the poles. If rotation of the loop or loops is continued another 180°, it will induce electricity in the opposite direction. Under this condition the current in the loop would be alternating in direction; if a commutator is added so that the current that is taken off is traveling in one direction, the tapped off current is called "direct." If the current is not changed by such a commutator in direction, then the current will be alternating. The principal advantage of the alternating current is that it can be changed in voltage by the transformer.

The electric motor efficiency in converting the electric energy into mechanical energy may be approximately 90%, thus surpassing both the gas engine and the steam engine as an energy converter. The electric motor is easy to start and stop and with designed instrument can be controlled automatically. Most motors are designed for an overload capacity. Their limit of overload is determined by their heating-up capacity when in service, maintaining a temperature below some design critical point.

Electric motors are available for most industrial conditions, even to the driving of pumps under water. One motor with automatic controls can often replace 1 or more men and can serve 24 hr a day with dependability.

LICENSE TO INSTALL ELECTRIC WIRING AND EQUIPMENT

Many states and some larger cities have laws governing the engagement in business of electrical wiring installations. This generally does not apply to the individual who may want to wire his own premises without a permit. Some state fire insurance organizations may require that all wiring on an insured building must meet the National Electrical

Code and will require an authorized inspector to pass upon all wiring done.

BIBLIOGRAPHY

ANON. 1969. Electric Wiring Illustrated. Montgomery Ward Co., Chicago, Ill.

BODINE ELECTRIC CO. 1959. Bodine Electric Motor Handbook. Chicago, Ill.

BREDAHL, A. C. 1957. Wiring Manual. McGraw-Hill Book Co., New York.

ELENBAAS, W. 1959. Fluorescent Lamps. Macmillan Co., New York.

HEUMAN, G. W. 1959. Electric Motors. John Wiley & Sons, New York.

LIBBY, C. C. 1960. Electric Motors and Selection. McGraw-Hill Book Co., New York.

LUCKIESH, M. 1924. Light and Work. D. Van Nostrand Co., Princeton, N.J.

MAGNUSSEN, C. E. 1929. Direct Currents. McGraw-Hill Book Co., New York.

RICHTER, H. P. 1968. Electric Wiring Simplified. Park Publishing Co., Minneapolis, Minn.

SAY, M. G., and PINK, E. N. 1943. Synchronous Motors and Alternating Current Machines. Isaac Pitman and Sons, London.

CHAPTER 16

Prime Movers for Refrigerated Storages

COLD STORAGE POWER

Prior to World War I, most cold storage plants were powered by steam engines. There were 2 reasons for this: (1) fully 70% of cold storage warehouses were associated with ice plants and a major portion of ice plants were dependent upon steam condensate for their "so-called" distilled water for clear ice; (2) condensing steam turbines that might well supply ample condensate for the ice-making function were not adapted to operating at the low speeds essential to driving refrigerant compressors. Even if condensing turbines might be geared down to drive refrigerant compressors, equipment of this type was not available on the American market before 1925.

From 1885 to 1915, most large refrigerant compressors were driven by horizontal compound Corliss engines. The compressors were of the slow speed, long stroke, small bore design often of the jack knife type; that is, the steam engine was horizontal, the compressor vertical, primarily to economize on floor space.

The change over to electric or diesel engine drive of the compressor came between 1920 and 1935. This was brought about by: (a) change over from distilled to raw water for ice making; (b) adoption of high speed, short stroke refrigerant compressors for both ice making and cold storage; and (c) more attractive power rates for purchasers who would use synchronous motor drives for all applications requiring an electrically driven compressor of 50 hp or more. Such a synchronous motor was most helpful to the power company in establishing a better power factor in the region of the ice plant or warehouse.

There are a few of the older ice and cold storage plants that are still using condensing Corliss engines and making distilled water ice, especially where the available tap water is high in gypsum or in common salt. Some post World War II cold and freezer warehouses of large size have turned to gas turbines to produce their own electric power. This usually is done to make available two sources of power at the switchboard, one of which is designated as standby in case of a blackout of either source.

194

Steam Engines

Unfortunately, the size of nearly all steam-driven cold storage warehouses is comparatively small. This makes it advisable when steam engines are used, to install low cost, noncondensing, units in most plants. For such units the performance economy is relatively low, but the simplicity of operation and low maintenance expenditures of the low-speed, noncondensing Corliss drive makes it a favorite with both manufacturers and users, especially when ample boiler capacity is available to furnish steam for the engines anticipated.

With the long-stroke, slow-speed, steam-driven units, high volumetric efficiencies were readily obtainable. The fundamental equation for net volumetric efficiency of ammonia compressor is made up to two component efficiency factors, which in equation form reads

$$e_v = e_s = e_c$$

where e_v = net volumetric efficiency

e_s = volumetric efficiency computed on the basis of effective superheating during compression

e_c = volumetric efficiency computed on the basis of the clearance

Interpreted, this formula states that the net volumetric efficiency is equal to the product of the volumetric efficiency due to superheating and the volumetric efficiency due to clearance. The superheating effect is usually greater at high speeds than at low speeds, and this gives higher values of e_s for slow-speed machines.

The ratio of the clearance volume to the total cylinder volume is greater on short-stroke than on long-stroke machines, and this gives a higher value to e_c on the long-stroke machine. Therefore, any attempt to change from slow-speed steam driven units to the high-speed, electric driven units has to be made with some sacrifice in net volumetric efficiency of the refrigeration machine, as both e_s and e_c are lowered in value. Fortunately for the designers of high-speed ammonia compressors, vertical single-acting machines can be used. The vertical single-acting machine has always offered greater possibility for high volumetric efficiencies than the horizontal double-acting ammonia compressor. Vertical machines could be made with false heads, and thus permit very small clearance, but the horizontal double-acting machine could not be easily designed to meet these conditions.

Although designers of steam-driven machines have met this situation by using vertical compressors with horizontal engines, it has involved an expensive construction. Thus, while the vertical slow-speed, single-acting, steam-driven ammonia compressor has given high volumetric efficiencies, the machine cost has been high, and usually the steam economy relatively low. Its electrical competitor has had to sacrifice something in both volumetric efficiency and length of life, but the cost of equipment and space required have been greatly reduced.

Internal Combustion Engine Drives for Refrigeration Plants

For cold storage, ice production, locker storage and general refrigeration plants the Otto and Diesel cycle oil engines are most serviceable as a standby unit. In some of the smaller cities that are removed from the large electrical distribution systems, and even in some cities where electrical power is available but is sold at prohibitive rates, either the Otto or Diesel engine may be found to be most economical as the principal engine and as well for standby service.

In accepting the responsibility for providing uniform low temperature for the storage of perishable products, one of the first requirements for a successful storage plant is dependable power. Where this cannot be guaranteed by the electrical suppliers, it is recommended that a stand-by emergency power supply be furnished by either Diesel or other engine power. If the electrical power is not dependable, it is advisable to install both the principal power source and the stand-by of Diesel or other oil power.

For the smaller installations such as locker plants, the gasoline engine with full automatic controls is most desirable. Diesel engines are not generally available in capacities below 30 hp.

Classification of Internal Combustion Engines.—Internal combustion engines may be classified by several methods. Classification may be by the fuel burned, such as kerosene, butane, gasoline, gas or oil; by cylinder arrangement, as horizontal, radial, vertical, opposed, etc; by type of cycle, such as Otto, Diesel, etc; by the number of cylinders, as single cylinder, multi-cylinder; and by application, such as tractor, automobile, airplane, marine, stationary, etc.

The internal combustion engine may be of low, medium or high pressure. The low-pressure engines are those that generally do not exceed compressions of 150 lb and an explosion pressure of 250 lb. The injected fuel is sprayed on to a hot bulb or a hot plate where it is heated and vaporized; with the heat of compression ignites the fuel-air mixture and performs the work of the engine power stroke.

The medium-pressure engines develop a compression pressure of 250 to 350 psi, then the fuel is injected at the end of the power stroke. The explosion will raise the pressure to 400 to 500 psi.

The high-pressure, more commonly called Diesel engine, may be either 4-stroke or a 2-stroke cycle engine. The charge of air is compressed in the cylinder to approximately 500 psi which results in a temperature of probably 1,000°F. At the end of this stroke the fuel oil is fed into this hot compressed air by a high-pressure pump, and a slow ignition results.

Turbines

By definition the turbine is a machine for generating mechanical power in rotary motion from the energy in a stream of fluid. The energy in the fluid, originally in the form of head or pressure energy, is converted to velocity energy in passing through a system of stationary and moving blades in the turbine. Changes in the magnitude and direction of the fluid velocity are made to cause tangential forces on the rotor blades, thus producing power as the rotor turns. The most commonly used fluids for turbines are steam, hot air or gaseous products of combustion, and water. Turbines drive about 95% of all electrical power producing generators in the world, and are used for innumerable other mechanical drives.

Steam Turbines.—Steam turbines came into use in North America about 1910. For most refrigerant compressors of that date the speed of steam turbines was too high for practical application to the existing low speed refrigerant compressors. When steam turbines were used for power for cold and freezer storage energy, whether privately financed or through a public utility, the turbine was directly connected to an electric generator and the power at the refrigerant compressor was supplied through an electric motor.

Steam Turbine Classification.—When classified according to method of steam expansion a turbine is specified as (a) impulse, or (b) reaction. In the impulse turbine the fluid is accelerated through the stationary blades and then made to impinge upon the rotor blades or buckets, thus having its direction changed and producing a force on the rotor with substantially no pressure change while passing through the rotor.

Impulse Turbine.—In an impulse turbine the expansion occurs in the stationary blades or nozzles only. By this expansion the steam gains high velocity and impinges against the blades or buckets and drives them forward, causing the rotor to revolve.

Reaction Turbines.—In the reaction turbine the expansion of the steam occurs in the rotating blades or buckets. The reacting force of this steam gaining velocity in the blading is similar to that experienced when trying to hold a fire hose under high pressure. The hose reacts or swings back with great force.

Steam Turbine Classification by Stages of Expansion.—When classified according to stages, steam turbines are made up as (a) single stage, and (b) multi-stage. Steam, when permitted to expand from a high pressure through a nozzle to a low pressure, may reach a velocity of 3,000 or 4,000 ft per sec depending upon the initial pressure of the steam and the back pressure against which it is expanding. For the greatest efficiency in the simple impulse turbine, the blades or buckets driven by the steam should travel at somewhat less than ½ the speed the steam would travel if no obstruction were placed in its path. When the turbine is so constructed that the steam passes through and has its entire pressure drop in just one set of nozzles or blades in going through the turbine, it is classified as single stage.

In the reaction turbine, about half of the available energy is used to produce a velocity through the stationary blades approximately equal to the rotor velocity, and the remainder of the pressure energy then accelerates the fluid while it is passing through the rotor, thus producing a drive force from the acceleration.

The rotative speed of such a unit is necessarily very high. Some single stage turbines have been built with speeds of 24,000 to 30,000 rpm; common practice, however, is from 3,000 to 8,000 rpm.

By making the pressure drop in steps requiring the steam to pass through a nozzle or set of nozzles, then through a stage of blading; then through a second set of nozzles or stationary blades and through a second set of blading attached to the same shaft as the first set, and continuing this pressure step-down process in successive stages, it gives the effect of reducing the velocity of each stage but gives the greater number of pounds of rotative pull on the shaft. This is known as multi-staging. It enables the unit to travel slower but to pull harder.

The multi-stage turbine gives the same horsepower at a slower speed than the single-stage turbine for the same amount of steam used. Multi-staging is, then, the designer's method of getting slower speeds on turbines.

Gas Turbines.—Gas turbines came into commercial use as a power source after World War II. They are available for many applications in gas turbine electric generator sets and for aero, marine and automobile power units.

Gas Turbine Arrangements.—Gas turbines can be constructed for single or multiple-shaft arrangements. They can be arranged to supply power, high-pressure air, or hot-exhaust gases either single or in combination.

A single-shaft unit, consisting of a compressor, combustor and turbine, is a compact, lightweight power plant. It is capable of rapid starting and loading and has no standby losses. In can be arranged to use little or no cooling water, which makes it particularly attractive for powering transportation equipment and as a mobile standby and emergency power plant. This type of plant can compete efficiently with small steam plants. Its simplicity involves a minimum of station operating personnel; some of these plants are arranged for completely remote operation.

To improve efficiency, the energy in the exhaust gases can be used either in a waste-heat boiler, or combination with other processes. For example, a unit arranged to supply process air at 35 psig for operation of blast furnace can use blast-furnace gas for its fuel.

Gas Turbine Fuels.—In the open-cycle plant, products of combustion come in direct contact with the turbine blades and heat-exchanger surfaces. This requires a fuel in which the products of combustion are relatively free of corrosive ash and of residual solids that could erode or deposit on the engine surfaces. Natural gas, refinery gas, blast-furnace gas, and distillate oil have proved to be ideal fuels for open-cycle gas turbines. Combustion chamber and fuel nozzle maintenance are neglible with these types of fuel. Residual fuel, oil treated to avoid hot ash corrosion and deposition is also satisfactory although it requires frequent cleaning of fuel nozzles and higher combustion maintenance. Vanadium pentoxide and sodium sulfate are the principal ashes that have been found to cause corrosion and deposition in the $1,2\overline{5}0°F$ to $1,600°F$ temperature ranges of modern day gas turbines.

As a more recent development in the gas turbine list of fuels, there is an increasing interest in using powdered coal; the Swiss have been unusually successful in their experimental tests of powdered coal driven gas turbines. The Brown-Boveri Company of Switzerland have made available experimental units of this type with considerable success.

Turbojet Engines.—The turbojet engine is of the gas turbine type. The thrust or push of a turbojet engine comes from reactions set up inside the engine as the gases are accelerated through it and out the exhaust nozzle. It has not invaded the field of refrigeration except in air cooling of moving vehicles. The propulsive ability of a turbojet engine is measured in pounds of thrust rather than in horsepower (which is used for piston and turboprop engines) because no actual engine motion is involved when the aircraft is standing still. Once the airplane starts to move, however, a comparison between thrust and horsepower can be made, with the following equation

$\text{T hp} = \text{TS}/375$ where T hp is thrust horsepower
T is thrust in lb, and
S is speed in mph

Thus, at 375 mph, 1 lb of thrust equals 1 thrust horsepower.

Turbojet engines are currently only applicable to cold and freezer warehousing when such storage is carried on in moving vehicles such as: refrigerated trucks, marine transport on relatively small ships, airplane transport and for specialized transportation of frozen foods. With the development of the turbojet power units to high speed transportation of any of these fields, new developments can be anticipated even to the suggested use of turbojet freight airplanes that permit freezing of produce by interstellar air at high altitudes enroute from producing to consuming centers.

TABLE 16.1

TURBOPROP AND TURBOJET ENGINES CLASSIFIED BY SPEED RANGE

Type	Speed Range, Mph
Turboprop	0–500
Single-shaft, single-stage centrifugal flow	
Single-shaft, 2-stage centrifugal flow	
Single-shaft axial flow	
Twin-shaft twin-spool axial flow, with free turbine	
Turbojet	500–750
Single-spool axial flow	
Twin-spool axial flow	
Twin-spool axial-flow by-pass	
Twin-spool axial-flow by-pass, with turbofan	
High pressure ratio turbojet	750–1500+
Single-spool axial flow	
Twin-spool axial flow	
Turbojet with progressively diminishing pressure ratios	100–1700+
Single-spool axial flow	
High-temperature types	

ELECTRIC MOTOR REFRIGERATION DRIVES

Advantages offered by electric drive compressors are: (1) cost less to equip plant per ton of refrigeration; (2) occupy less space; (3) can be operated with less plant labor; (4) operating expense is nearly proportional to output; and (5) greater flexibility in operation.

The disadvantages are: (1) high speed results in lower volumetric efficiency of compressor; and (2) high speed involves more repair expense and shorter life of equipment.

Electric Motor Selection

The first step in the selection of motor drives is to decide between direct and alternating current as the power source. In many localities this decision will be made readily if power is to be purchased, since the available power will be the most economical.

If power is to be generated by the owner, then the three-phase electric drive usually will be the most satisfactory. The cost of generation, the simplicity of distribution and the cost of electric motors of the alternating type are in favor of three-phase ac power adoption.

Single-phase motors are not only more expensive but they are more subject to repair than three-phase motors. The cost of power distribution of single-phase motors is also greater than for three-phase frequencies.

In ac cycles or frequencies, 60 cycles, or hertz, is the most common within the United States and Canada, although 25-hertz current is available in several large cities. Usually the supply of available motors in the stores of the North American area is predominantly 60-hertz 3-phase. In England and Europe 50-hertz current prevails.

POWER CONSUMPTION COMPUTATIONS FOR DIFFERENT TYPES OF MOTORS

Direct-Current Power

While relatively few dc motor drives are installed today because of the predominance of ac central station power, conditions occasionally require a dc installation. In using dc power, the current consumption can be computed directly from the ammeter and voltmeter on the switchboard by multiplying amperes by the volts

$$\text{volts} \times \text{amperes} = \text{watts, or } E \times I = P$$

where E = electromotive force (volts)
I = amperes
P = power in watts

The power in watts can be readily changed to kilowatts by dividing the watts by 1,000, or

$$\frac{P \text{ (in watts)}}{1,000} = P \text{ (in kilowatts)}$$

Since most rates are made on the basis of kilowatt-hours, it will be necessary to multipy the reading in kilowatts by the time in hours to change to kilowatt hours.

$$P \text{ (in kilowatts)} \times T = P \text{ (in kilowatt-hours)}$$
where T = time in hours

Single-Phase AC Power

In measuring single-phase ac power, a new term is introduced as compared with the calculations in dc power. This new term is the power factor.

In measuring dc power, the power used was determined by multiplying the amperes by the volts. This same procedure would be correct for single-phase alternating current, but the voltmeter and ammeter do not always indicate the true effective volts and amperes. This is caused by the inductive and the capacitive effect in an ac system. The power formula for a single-phase ac system will therefore be

$$E \times I \times pf = P \text{ (in watts)}$$

Power Factor and Load Factor

Engineers often confuse the term power factor with load factor and use them indiscriminately. The value of the power factor may vary from 40% or 0.40 under very bad conditions to near 100% or 1.00 under more ideal operating conditions. The low power factor does not indicate that the efficiency is appreciably lower than that at high power factor. The actual power purchased is based on the true kilowatts. But low power factor does introduce a bad operating condition by excessive heating and should be corrected if possible.

Load factor, as usually computed, is a ratio of the average kilowatts used over a stated period of time to the maximum demand as based on a maximum-demand meter reading. The period of time on which this maximum is based is important. A period so short as when an unusual rush of current caused by some very temporary disarrangement becomes the basis of computation would be manifestly unfair. Such a maximum is generally based on an average of the three highest loads maintained over a 5-, 10-, or 15-min period during the month.

Problem 1.—The average voltmeter reading on a single-phase motor is 222 v. The average current reading is 91 amp. The power-factor meter indicates a power factor of 81%. How much power is being used?

Solution. For single-phase circuits the power consumption is

$$E \times I \times pf = P \text{ (in watts)}$$

in the above problem

$$E = 222 \text{ v}$$
$$I = 91 \text{ amp}$$
$$pf = 0.81$$

Then $222 \times 91 \times 0.81 = 16{,}363$ watts
The power in kilowatts will be

$$\frac{16{,}363}{1{,}000} \text{ or } 16.353 \text{ kw}$$

Problem 2.—The single-phase voltage of a circuit is 440. The current reading is 200 amp. The power-factor meter shows 55%. What is the indicated power load in kilowatts?

Solution

$$E \times I \times pf = P \text{ (in watts)}$$

Then $44 \times 200 \times 0.55 = 48,000$ watts. The power consumption in kilowatts will be

$$\frac{48,000}{1,000} = 48.4 \text{ kw}$$

Three-Phase AC Power

Most refrigeration compressors are driven by three-phase motors. Two general types are available. These are induction motors and synchronous motors. Investment cost is in favor of induction motors.

Synchronous motors are to be preferred for most three-phase installations. The improvement which they give to power-factor conditions is very desirable to both the owner and to the company selling power. It is to their mutual interest to use synchronous drives.

The power equation for computing the power consumption from the voltmeter and ammeter reading of a 3-phase circuit involves an additional term because of the 3 circuits involved.

TABLE 16.2

WIRE AND CURRENT TABLE FOR DC MOTORS

Horse-power	110 Volts		200 Volts		440 Volts	
	Average Amps at Full Load	Size of Wire Rubber Insulation	Average Amp at Full Load	Size of Wire Rubber Insulation	Average Amp at Full Load	Size of Wire Rubber Insulation
1	9	14	4.5	14	2.3	14
2	16	12	8	14	4	14
3	24	8	12	14	6	14
5	40	6	20	10	10	14
7.5	60	4	30	8	15	12
10	75	2	38	6	19	10
15	112	0	56	4	28	8
20	148	000	74	2	37	6
25	184	0000	92	1	46	5
30	220	250,000[1]	110	0	55	4
40	290	400,000[1]	145	000	73	3
50	360	500,000[1]	180	0000	90	1
60	—	—	220	250,000[1]	110	00
75	—	—	275	350,000[1]	138	000
100	—	—	365	500,000[1]	183	0000
125	—	—	450	700,000[1]	225	250,000[1]
150	—	—	550	900,000[1]	275	400,000[1]

[1] Circular mils.

TABLE 16.3
TABLE 16.3

WIRE AND CURRENT TABLE FOR THREE-PHASE AC MOTORS

	110 Volts		220 Volts		440 Volts	
Horse-power	Average Amp at Full Load	Size of Wire	Average Amp at Full Load	Size of Wire	Average Amp at Full Load	Size of Wire
0.5	4	14	2	14	1	14
1	7	12	3.5	12	2	14
2	12	8	6.0	12	3	14
3	18	6	9	10	4.5	14
5	30	6	15	10	7.5	12
7.5	42	5	21	8	10.5	12
10	58	3	29	6	14.5	8
15	80	1	40	4	20	8
20	110	00	55	3	27.5	6
25	122	000	66	1	33	5
35	—	—	86	0	43	4
50	—	—	130	000	65	1
75	—	—	190	300,000[1]	95	0
100	—	—	250	500,000[1]	225	000
150	—	—	365	600,000[1]	183	300,000[1]
200	—	—	475	800,000[1]	238	400,000[1]

[1] Circular mils.

The fundamental three-phase power equation is

$$E \times I \times pf \times 1.73 = P \text{ (in watts)}$$

The value of E (volts) and I (amperes) may be taken as the reading on any one phase on the assumption that the load is balanced on all phases. The constant 1.73 is derived from the square root of 3, this being the mathematical relation between phases.

Problem 1.—The balanced load on a 3-phase circuit shows 220 v per phase and 119 amp on each leg. The power factor is 96%. What is the power consumption in 12 hr?

Solution.—The formula for 3-phase circuits is

$$E \times I \times pf \times 1.73 = P \text{ (in watts)}$$

Since
$$E = 220 \text{ v}$$
$$I = 119 \text{ amp}$$
$$pf = 0.96$$

then $220 \times 119 \times 0.96 \times 1.73 = 43,479.7 \text{ watts}$

To reduce this to kilowatt hours over a 12-hr period, we will have

$$\frac{43,479.7 \times 12}{1,000} = 521.76 \text{ kwh}$$

Problem 2.—The voltage of a 3-phase induction motor is 220 on each phase, with 63 amp flowing in each phase. The power-factor meter indicates 90%. What power in watts is indicated?

Solution.—For 3-phase circuits,

$$E \times I \times pf \times 1.73 = P \text{ (in watts)}$$

Since

$$E = 220 \text{ v}$$
$$I = 63 \text{ amp}$$
$$pf = 0.90$$

then

$$220 \times 63 \times 0.90 \times 1.73 = 21,580 \text{ watts}$$

FOUNDATIONS FOR POWER UNITS

The theory of foundation design is complex. To cover all types of power units would require several chapters. Specifically the foundation should become an integral part of the earth and the engine or turbine should be rigidly fastened to this foundation. No felt pads or rubber supports should be used between the foundation and the power unit to absorb shock vibration.

By making the foundation an integral part of the earth any vibration received will be transferred to the earth; the mass of the earth is so large and that of the engine or turbine so relatively small that any unbalanced force within the unit will be practically absorbed by the combined mass of the engine and of the foundation of the earth itself.

The weight distributed over the mass of the foundation due to the combined weight of such foundation and the engine or turbine installed should not be such that it will cause the earth beneath to settle or flow away. The bearing strength of the earth may vary from almost nothing in swampy quicksand to 100 tons per square foot in uniform quartzite or granite.

The foundation materials used may be: rock, brick, wooden piles and concrete with steel reinforcing; however, current practice is to make all foundations of a reinforced concrete mixture.

Foundations for Turbines

The foundations for turbines are usually reinforced concrete arranged to permit the condensers or other related auxiliaries to be placed directly under the prime mover to conserve space. The alignment and adjustment of turbines must be made with great care and most suppliers of turbines will require the plant engineer or consultant of the purchaser to submit

TABLE 16.4

SAFE BEARING LOADS OF SOILS AND ROCKS

Kind of Material Beneath Foundation	Number of Tons of Load Such Ground Will Carry per Sq Ft
	Tons
Extra hard rock without faulty seams	100
Soft rock and seamy rock	15
Shale rock	5
Clay and flint gravel	6
Hard clay	4
Common clay	2
Gumbo or plastic clay	½ to 1
Dry sand and clay	4
Dry sand	2
Marsh sand	¼ to ½

a preliminary foundation designed for supplier approval then guaranteeing the unit to operate without noticeable vibration. The supplying contractor of the power unit then specifies the final acceptable design and usually helps to supervise the foundation building and unit erection up to the completion of the trial tests and contract.

Foundation Materials.—Rock, brick, wooden piles and concrete have all been successfully used for foundations. Today practically all foundations are made of concrete.

Concrete Mixtures Used.—Mixture 1: One part cement, 3 parts sand and 6 parts clean stone 2.5 in. diam or less.

Mixture 2: One part cement, 2.5 parts sand, and 5 parts clean stone 2.5 in. diam or less.

Mixture 3: One part cement, 2 parts sand, and 4 parts clean stone 2.5 in. diam or less.

Mixture No. 1 is the cheapest. It is satisfactory where clean sand and rock are available, and where the foundation is comparatively large.

Mixture No. 2 is recommended for small foundations, or in large foundations carrying heavy concentrated pressures. Mixture No. 2 is also used in large foundations carrying relatively small loads, but in which a low grade of sand must be used.

Mixture No. 3 is desirable where extra strength is required. For wall foundations, out bearing piers, and similar heavily loaded locations it is desirable.

Amounts of Each Kind of Material to Purchase

Table 16.5 is made out to show the amounts of materials required for the different mixtures. Cement is sold in either bags or barrels. There

TABLE 16.5

AMOUNT OF DIFFERENT MATERIALS REQUIRED TO MIX ONE CUBIC YARD OF CONCRETE
WITH DIFFERENT MIXTURES

Mixture Proportions	Quantity of Portland Cement Required Bbl	Quantity of Sand Required Cu Yd	Quantity of Rock Required Cu Yd
1 cement, 3 sand, 6 rock	1.0	0.46	0.92
1 cement, 2.5 sand, 5 rock	1.2	0.46	0.92
1 cement, 2 sand, 4 rock	1.5	0.45	0.90

An average cubic yard of dry concrete will weigh 2 tons.
An average cubic yard of dry sand will weigh 1.5 tons.
An average cubic yard of 2.5-inch stone will weigh 1.5 tons.
A barrel of cement is approximately 4 cu ft or 376 lbs.
A sack of cement is approximately 1 cu ft, or 94 lbs.

are four bags of cement to the barrel. Sand and stone are usually sold by the cubic yard. There are 27 cu ft in a cubic yard.

Cold storage plants utilizing commercial power are usually supplied with power at voltages of 220 v or higher up to 550 v with the most common voltage being 440 or 480 v. A 480-v 3-phase system is usually supplied in the form of a 4 wire, Y connected system with 480 v between phases and 277 v from any phase to the neutral or ground. With this type of system, all components are standard and readily available without premium costs. Wire sizes are not excessive and most electricians are familiar with this voltage and installation presents no unusual problems.

Power is normally supplied by the power company to the cold storage facility through a bank of transformers serving only the cold storage plant and normally at a single voltage. Wherever possible, two incoming sources of power are desirable, connected to the transformer banks through an automatic transfer switch. If power failure occurs on the high line normally serving the transformer bank, the second source is automatically switched to feed the bank and the primary source cut out. On resumption of power to the primary source, it automatically resumes service to the plant. This type of protection tends to prevent any prolonged outages of power.

The incoming power from the transformer bank will normally connect directly to a load center. The load center may be of the factory fabricated type custom-built for a particular plant or it may be made up of standard purchased components. In the load center will be the main incoming switches. Smaller plants may have one main incoming switch with a capacity up to 2,000 amp. As plant size increases, up to six main switches may normally be used to divide the load. This should always be checked against local codes as to the number of switches allowed. Less than the maximum number of main switches should be used, when possible, to allow for future growth. All main entrance switches, particularly when served by large network power sources, should be protected by high interrupting capacity fuses made especially for this type of service. Sizes in individual plants can vary greatly from the value above. The size of switches to be used with the attendant busway connections should be checked to obtain the least expensive combination.

Each main entrance switch will normally be connected to a number of separate feeder switches by busway connections in the load center. The feeder switches normally fall in the size range of from 200 to 600 amp and from these switches, the various components of the plant electrical system are fed. These components consist of electric motors, control apparatus, lighting, battery chargers and all other electrical equipment installed in the plant.

When the plant load center is located in the machine room in an automated plant, the automatic control panels are quite often made a part of the main switchgear in that the cubicles containing the control circuits and electrical switching relays adjoin the power cubicles to make a continuous panel system. Sometimes gages and indicating instruments etc. are also added to the control panels as well as indicating alarm systems so that the load center actually contains the entire electrical system center for the entire plant. Control wires for compressors may be run in the conduits carrying power to the compressor motors to pick up the various safety and operating devices located at the compressor and driving motor assembly.

Most designers find it desirable to use a voltage different from the power voltage for the control circuits with 115 to 120 v being a popular choice. A control transformer can be mounted in one of the control cubicles to supply control power at the desired voltage for the entire refrigeration system control. By utilizing this single source of control power through a magnetic contactor, a desirable safety feature of shut off may be made available. By means of a control toggle or other type switch mounted at the entrance to the machine room, the entire control system may be shut off without entering the machine room. With fail-safe controls normally used in automatic control systems, the entire plant can be shut down without entering the machine room. This is a very desirable feature in case of fire or a bad refrigerant leak.

From the central panel, or load center, power is normally distributed to the machine room and to the entire plant. Various circuits are run to wherever is necessary in the plant to supply power and lights in the various areas. As a general rule, all circuits are run from the load center at the plant voltage to the various locations. If a different voltage than the plant voltage is required at a particular location, dry type transformers are used to supply the divergent voltage. Battery chargers and integral horsepower motors will usually be operated at plant voltage in the 480-v supply system. Fractional horsepower single phase motor loads and incandescent lighting will normally require a 115- or 120-v circuit. From a 480-v plant system, dry type 3-phase transformers may be used to supply 208-v, 3-phase, 4-wire service for both single and 3-phase loads. These transformers will most often be sized from 25 kva to 75 kva depending on the load to be served.

Lighting in cold storage warehouses is not particularly standardized and many different types and densities of lighting will be found. Fluorescent lighting gives very good results for dry storage, cooler and freezer storage. Special jacketed tubes are used in cold temperatures and with this type of tube, lighting intensity is very good. Fluorescent lighting may be installed using 277-v circuits obtained from a phase to neutral in a 4-wire, 480-v system. This eliminates the need for localized transformers for lighting in 480-v plant electric systems. Lighting circuits are handled in a conventional manner except the circuit breakers are rated at 277 v instead of the normal 115 v. Two 8 ft fluorescent tubes mounted end to end with a single ballast makes a good unit for storage rooms. These 2 units will use about 450 watts of power for operation. Good lighting can be obtained by utilizing two 8-ft tubes for approximately 1,000 sq ft of floor area. This will give adequate light for most standard warehouse operations, even with the light source as high as 20 ft above the floor. On docks, the two 8-ft lights can be mounted side by side at about 50-ft intervals for general dock illumination. No special reflectors are required for this type of illumination unless desired by the owner to enhance further the lighting. The above is a good level of general lighting. Some owners may desire higher lighting densities for various reasons.

Mercury vapor lights are used quite frequently for outside security lighting. These lights can also be obtained for 277-v operation and fed from the lighting panel board. Outside security lights are quite frequently controlled from a time clock which is reset periodically to conform to the hours of darkness.

Blast freezer areas will normally require incandescent lighting since fluorescent lights are not satisfactory in extreme low freezer temperatures, particularly where air blast is present. Incandescent lighting of one type or another is usually required, to some extent, in many areas of the plant. The lighting circuits for incandescent lighting are usually obtained from the step down transformers located in the various areas of the plant. Leased spaces will also require special attention and transformers for the power peculiar to the type of equipment in the space. Quite often 208- or 220-v 3-phase as well as single-phase power will need to be available for special processing equipment.

All electrical wiring in the plant should conform to the national and local electrical codes for safe wiring. In designing an electrical system, it should be remembered that the electrical codes are based normally on the minimum safe installation and are not concerned with maximum installations. When installing an electrical system, the sizing of main and branch circuits should always be done generously. All plants, if successful, tend to grow over the years and wiring should never be put in for minimum use. Where warehouses cover large areas, supply electrical lines in the plant should always be checked for voltage drop as well as load carrying ability. Quite often it will be found that long supply lines will need to be considerably larger than required for current carrying capacity to prevent too high a voltage drop. In running long power lines it is advisable to check out cost of multiple conductors over a single conductor per phase. Most codes will allow the use of conductors in parallel when certain installation rules are followed. Multiple conductors will quite frequently show a considerable savings over a single conductor of equal current carrying capacity.

Extra care should be used when running electrical connections through insulated walls into a cold area. Careful seal-offs with good sealing compounds is required to prevent trouble at some future date. This is particularly true in freezers.

Power factor correction will sometimes be required in plants utilizing a large number of small fractional horsepower electric motors. This can be rather easily accomplished by the use of a capacitor across the terminals of the motor. These capacitors will usually be of a production type that are relatively low in first cost; correction is applied at the motor where needed and the resultant reduction in current in the lines can save in wiring costs.

The layout of a complete electrical system for a cold storage plant is quite complicated. A great multiplicity of circuits is required, particularly in the modern automated plant. Careful preplanning of the system with generous wire sizing can save much grief and expensive rewiring.

BIBLIOGRAPHY

Anon. 1951. Standard practices for low and medium speed stationary Diesel engines. Diesel Engine Mfr. Assoc. Chicago, Ill.

BOYER, G. C. 1943. Diesel and Gas Engine Power Plants. McGraw-Hill Book Co., New York.

JENNINGS, B. H. 1963. Energy for Refrigeration, Progressive Refrig., Vol. 1. Intern. Inst. of Refrigeration, Paris.

JENNINGS, B. H., and ROGERS, W. L. 1953. Gas turbine analysis and practice. McGraw-Hill Book Co., New York.

KEARTON, W. J. 1948. Steam Turbine Theory and Practice. Pitman Publishing Corp., New York.

KENT, R. T. 1950. Mechanical Engineers' Handbook. John Wiley & Sons, New York.

LICHTY, L. C. 1951. Internal Combustion Engines. McGraw-Hill Book Co., New York.

MARKS, L. S. 1951. Mechanical Engineers' Handbook. McGraw-Hill Book Co., New York.

MEYER, A. 1939. The combustion gas turbine. Proc. Inst. of Mech. Engrs. 41, 197–212.

MOYER, J. A. 1934. Power Plant Testing. McGraw-Hill Book Co., New York.

PEW, A. E., JR. 1945. Operating experience with the gas turbine. Steam Eng. 15, 212–215.

RETTALIATA, J. T. 1950. Gas Turbines Plant Engineering Handbook. McGraw-Hill Book Co., New York.

SALISBURY, J. K. 1950. Steam Turbines and Their Cycles. John Wiley & Sons, New York.

SHOOP, C. F., and TUVE, G. L. 1956. Mechanical Engineering Practice. McGraw-Hill Book Co., New York.

SORENSEN, H. A. 1951. Gas Turbines. The Ronald Press Co., New York.

STANIAR, W. 1950. Plant Engineering Handbook. McGraw-Hill Book Co., New York.

STODOLA, A. 1927. Steam and Gas Turbine. McGraw-Hill Book Co., New York.

VINCENT, E. T. 1950. The Theory and Design of Gas Turbines and Jet Engines. McGraw-Hill Book Co., New York.

WOOLRICH, W. R. 1966. Handbook of Refrigerating Engineering, 4th Edition, Vol. 2. Avi Publishing Co., Westport, Conn.

Warehouse and Freezer Management and Use

CHAPTER 17

The Rise and Expansion of the Refrigerated Food Industry

INTRODUCTION—ORIGIN OF MEAT, FISH AND POULTRY FREEZING

Freezing has long been recognized as an excellent method of food preservation. Since early times, farmers, fishermen and trappers residing in regions having long cold winters have prepared their fish, game and other meat by freezing and storage in unheated buildings.

The freezing of beef and mutton was practiced in Midwestern United States and Canada even before these areas were divided into states and provinces. Farmers and ranchmen utilized snow and pond ice to pack their early winter killings of cattle and sheep for family and community use. These killings were timed to fit the advent of wintry or freezing weather and significant snowfall. These would reduce the hazard of meat spoilage before the proverbial late February thaw of some months later. To provide the home-killed meat needs for the year without requiring butcher shop purchasing, it was traditional to name the first frosty nights of fall "hog-killing weather." Thousands of hogs were slaughtered by individual families during this early fall chilly period. The most perishable parts of each hog, such as the heart, kidneys and liver were packed in pond or commercial ice and eaten as fresh. The other less perishable portions were converted into spiced and seasoned sausages and the rear and front quarters salted or smoked. Finally the side meat was rendered and leaner sections sliced and placed in jars of well-salted melted lard. These pork products provided probably 90% of the family summer meat consumption for the producing farm families, the other 10%, usually beef, mutton or fowl, being supplied on special days from their poultry flocks or from the butcher shop.

Actually frozen beef and mutton were accepted as most satisfactory meat provisioning by North American farm families many decades before frozen steaks and lamb chops were "tolerated" by the city gentry. Economic necessity of the farm and ranch meat producers of North America made frozen beef and mutton an acceptable table meat provision for many years before the more sophisticated tables of North America and Western Europe would admit that they could possess high quality and be delectable.

Australian and New Zealand Frozen Meat

The first successful shipment of frozen beef was transported by D. Peyton Howard from Indianola, Texas, to New Orleans, Louisiana, in 1867 and served in hospitals, hotels and restaurants. Simultaneously, Thaddeus S. C. Lowe equipped a vessel with commercial dry ice under British patent and transported frozen meat from New York to Louisiana.

The first successful shipment of frozen beef across the equator was carried on the steamship Paraguay from Buenos Aires to Marseille via Havre in 1877. This was followed by a successful shipment of frozen meat from Australia and New Zealand to the Smithfield Market, London. It originated at Melbourne, November 29, 1879 and arrived in London in excellent order, February 2, 1880.

By 1881, other steamships were equipped for freezer refrigeration of meat. The Dunedin sailed from New Zealand with 9,000 frozen carcasses of mutton and lamb and some butter for London. On arrival the product was advertised and sold as good and prime quality.

Advent of Quick Freezing

The advent of quick freezing revolutionized food freezing. The demand for frozen meats, poultry, eggs, fruits and vegetables gave new life to the industry. However, according to Enochian (1968), growth of the frozen fresh food industry was slow in its early years. Economic factors, such as the Great Depression of the 1930's, no doubt contributed to this slow takeoff. The time lag required to build up storage and distribution facilities and the lack of venture capital for investing in an industry whose future was still uncertain, also had a delaying influence on industry growth.

Consumer prejudices and long established habits had to be overcome, or had to await the fresh outlook of a new generation of Americans, before the industry could achieve a rapid rate of growth.

During the past 35 yr, frozen foods made tremendous gains due to the pioneering and continuing research by frozen food processors, container manufacturers, the state colleges, and state and federal government agencies. This research has been concerned with improvements in the freezing process itself, and improvements in the product through breeding of better strains, better selection, preparation, and handling of raw product. Equal attention is being given to preservation of product quality during marketing. An important part of the research has focused on developing information on the effects of time and temperature on frozen food quality during transportation, storage, and merchandising. As this information was translated into government and industry standards for maintaining good quality throughout the entire marketing process, as advertising and promotion attracted consumers, and as investments were made in facilities for handling frozen foods, the industry grew rapidly.

According to Anderson (1953), production of all frozen foods combined rose from 568 million pounds in 1941 to 1,028 million pounds in 1945, and then in 1946, it rose to 1,317 million pounds. The requirements of the armed services, shortages of the war and postwar periods, and derationing of frozen fruits and vegetables seven months sooner than canned goods, all contributed to the rapid expansion. In 1947, the boom came to an abrupt halt. Overproduction resulted in large inventories of frozen foods, some so poor in quality as to be inedible. Demand fell off, partly because of the restoration of normal conditions and partly because of the injury done consumer acceptance by the presence of

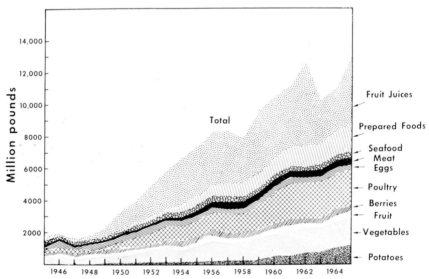

FIG. 17.1 GROWTH IN U.S. FROZEN FOOD PRODUCTION 1945–1965

Quantities based on best available industry and government sources.

inferior products on the market. In the squeeze that followed, many processors were forced out of business, as were most of the specialized frozen food stores and home-delivery services that had mushroomed after the war. The effect of those conditions on 1947 production is shown in Fig. 17.1. The setback was only temporary, for in the next year sales revived, and in 1949 production reached a new high.

Since 1949, production of frozen foods has grown steadily, reaching an all time high of nearly 13 billion pounds in 1962. This included 5.1 billion pounds of frozen fruit juices (reconstituted basis), most of which was concentrated orange juice which also reached an all time high that year. This surge in orange juice production was due to the exceptionally large crop of oranges in Florida that year. In 1963, the production of frozen foods fell below the level that had been achieved 5 yr earlier primarily because of the much smaller production of concentrated orange juice which was a result of a small orange crop that year. Since 1963 (Enochian 1968), there has been a resumption in the steady rise in total production of frozen food (Fig. 17.1).

THE ADVENT OF FROZEN PRECOOKED FOODS

Although the Birdseye Laboratories in 1932 tried to market pilot packs of a few frozen precooked foods (squash, corn on the cob, and lobster), it was not until near the end of World War II that a large line of these foods was packed. Since the end of that war, a very great number of frozen precooked entrées, vegetables, baked goods, desserts, meals on a tray (TV dinners), etc. have been offered. The quantity of the frozen French fried potatoes produced is greater than that of any other product. Many other precooked foods are frozen on a large scale. Frozen precooked foods have been generally accepted by modern homemakers as the best of convenience foods. The increase of production of these products is continuing steadily and is likely to continue.

Cost of Freezing

Contrary to what most persons believe, the freezing of foods is not a costly operation. In general, the cost of preparing fruits and vegetables for freezing is the same as the cost of preparing them for canning. There is not much difference in the cost of containers used for frozen foods and those for canned foods. In fact, in many instances the cost of containers for frozen foods is slightly less per pound of food packed. This is particularly true of the large containers used for products packed for the institutional and wholesale trade.

The relative costs of freezing and processing (pressure cooking) operations will depend upon three principal factors: (1) the length and severity of the process (cook) required to sterilize the product; (2) the cost of fuel for generating steam and the price of electricity; (3) the amount of labor involved in handling the products during freezing or processing. Certain fruits, e.g., grapefruit, and fruit juices, that may be sterilized by passing the can through a hot water bath or by flash pasteurization, may be processed for less than they can be frozen. On the other hand, unless the cost of steam is very low indeed, vegetables which require a long process (cook) may be frozen more cheaply than they can be processed. With the exception of locker plants and small freezers, the vast majority of the commercial freezing plants use ammonia machines. While there are slight differences in the efficiency of the various ammonia refrigeration machines, the difference in the cost of liquefying ammonia is not very great. Ice is commonly made by freezing purified water with ammonia refrigeration machines. Since ice commonly sells at wholesale for $6.00 to $10.00 a ton, and since vegetables contain, on the average 90% water and fruits about 85%, it is evident that the actual cost of refrigeration for the freezing of fruits and vegetables (other than labor) is not more than 0.3¢ per lb.

FUTURE EXPANSION OF FROZEN FOODS FROM THE TROPICAL ZONES

With the increase of population to many nations of the world, the needed food supplies, especially of fish and fruit, may well be augmented from the tropical seas and regions of the verdant earth. The food shortages of the Western Europeans and the British in the nineteenth century were finally solved by the frozen food importations largely from South of the Equator to supplement the established imports from North America. The temperate regions of Australia, New Zealand, Argentina, Uruguay, Chile, Southern Brazil and South Africa supplied a large amount of the perishable foods most acceptable to Western Europe by steamship. Each steamship in its 4 to 6 weeks journey with frozen meats and butter in the refrigerated holds, passed over nature's storehouse of billions of tons of tropical fish, exotic fruits and nutritious nuts not yet acceptable to the northern food brokers, chefs, green grocers, fishmongers and butchers of Europe and Britain.

The expansion of use of frozen foods that can be anticipated from the tropical and semitropical zones will require an entirely new organization of food suppliers to gather, freeze, transport under refrigeration, and market some of these billions of tons of fish from the equatorial seas, and nuts and fruits from the regions between 30° North and 30° South latitudes. Coincidentally, the natives of these same zones will learn the wisdom of conserving by freezing to prevent decaying and then marketing the surplus they now discard as inedible. This tradition in harvesting, transporting and marketing techniques of tropical perishable foods together with the expanding use of already established sources of frozen foods by those of lower income levels cannot be accurately predicted. Since all-round quality of most frozen foods, especially in flavor, color, texture and vitamin content, is superior to that of the canned product, it is only logical to assume that freezing will eventually become the most important method of food preservation in these areas.

TABLE 17.1

TRENDS IN PER CAPITA CONSUMPTION REVEALED IN FROZEN FOOD STATISTICS
FOR THE UNITED STATES—1963 vs 1965
(BY category)

Produce	1965 Lb	1963 Lb	2-Yr Consumption Increase in %
Vegetables	15.412	12.213	26.1
Poultry	11.703	11.024	6.24
Prepared foods	10.994	7.620	45.3
Juices and concentrates	9.259	7.307	26.7
Seafoods	4.816	3.545	35.7
Meats	3.919	2.821	38.9
Fruits	3.288	3.261	0.827
Total	59.389	47.791	

Table 17.1 is based on data presented by Messrs. Harold L. Franklin and Sam Martin and published in the February 1967 issue of *Quick Frozen Foods*. The net percentage increase in frozen food consumption in this 2-yr period was 17%. The greatest increase was in prepared foods, the smallest can be explained by the 26.7% increase in consumption of fruit juices and concentrates.

TABLE 17.2

ESTIMATED DOLLAR VALUE[1] OF ALL COMMERCIAL FROZEN FOODS,
UNITED STATES, 1940–1966
(In Millions of Dollars)

Year	Value
1940	108
1945	257
1950	500
1955	1,700
1960	3,037
1965	5,765
1966	6,245

Source: Anon. 1967G. *Quick Frozen Foods.*
[1] At conservative retail and average institutional prices.

TABLE 17.3

DOLLAR VALUE MAJOR FROZEN FOOD CATEGORIES 1959 TO 1968
(All Figures In Millions[1])

Year	Fruits	Vege-tables	Poultry	Meat	Sea-food	Pre-pared	Juices	Total[1]
1959	$160	$495	$609	$240	$282	$525	$438	$2,749
1960	150	529	757	266	351	527	455	3,037
1961	180	608	935	286	483	620	550	3,664
1962	174	594	948	347	529	714	654	3,960
1963	170	577	925	413	591	885	820	4,381
1964	190	693	1,002	579	734	1,198	850	5,246
1965	189	773	1,012	642	932	1,374	842	5,765
1966	269	905	1,176	703	1,031	1,502	663	6,250
1967	271	924	1,234	771	958	1,665	623	6,449
1968	261	1,033	1,206	809	1,139	1,927	657	7,033

[1] Loss of a point or two in rounding out figures may result in the categories not adding up to precise total.

TABLE 17.5

REFRIGERATED WAREHOUSE CAPACITY AT 0°F OR BELOW, 1939–67

Date October 1:	Public Warehouses 1,000 Cu Ft	Gross Space, 0°F or Below Private 1. Semiprivate 1,000 Cu Ft	Meat-packing 1,000 Cu Ft	Total 1,000 Cu Ft
1939	65,804	8,246	12,538	86,588
1941[1]	84,379	8,427	15,110	107,816
1943	86,245	9,500	14,973	110,718
1945	110,448	11,944	12,492	134,884
1947	116,957	15,473	13,017	145,447
1949	134,296	21,397	12,257	167,950
1951	169,388	29,142	14,180	212,710
1953	257,868[2]	347,328[2]	60,218[2]	29,242[2]
1955	281,865	73,897	20,739	376,501
1957	331,510	77,319	19,291	428,120
1959	350,068	89,059	16,644	455,771
1961	380,919	112,958	19,061	512,938
1963	421,003	137,593	16,774	575,370
1965	467,338	155,845	16,650	639,833
1967	519,182	177,267	15,341	711,790

[1] June 16, 1941.
[2] 19°F or below. U.S. Dept. Agr., Agr. Mktg. Serv., Agr. Est. Div.
(1) Does not include FF distributors, retailers, locker plants, military establishments or any warehouse holding food less than 30 days.

TABLE 17.4

FREEZER STORAGE SPACE CAPACITY

By Type of Warehouse

(1,000 Cu Ft)

	1967		1965		1963		1961	
	Gross Space	Net Piling Space	Gross Space	Net Piling Space	Gross Space	Net Piling Space	Gross Space	Net Piling Space
Public general	519,182	385,800	467,388	345,165	421,003	310,247	380,919	277,666
Private and semiprivate General	177,267	132,133	155,845	116,373	37,593	103,307	112,958	83,638
Meat packing	15,341	10,346	16,650	11,089	16,774	11,115	19,061	12,400
Apple houses, public	967	706	1,111	833	1,179	972	1,679	1,369
Private and semiprivate	2,791	2,006	2,471	2,033	2,662	2,252	1,709	1,473
Total	715,548	530,991	643,465	475,493	579,211	427,893	516,326	376,546

Gross space is total space inside refrigerated rooms measured from wall to wall and floor to ceiling. Net piling space is gross space minus the space provided for ventilation (outside of pile) and space occupied by coils, aisles, posts, sprinklers, etc.

BIBLIOGRAPHY

ADLER, C. 1967. Submersibles seek untapped supplies of fish for tomorrow's freezers. Quick Frozen Foods 29, No. 9, 139–144.

AMERICAN MEAT INSTITUTE FOUNDATION. 1960. The Science of Meat and Meat Products. W. H. Freeman Co., San Francisco.

ANDERSON, O. E. 1953. Refrigeration in America—A History of a New Technology and Its Impact. Princeton Univ. Press, Princeton, N.J.

ANON. 1891A. The frigerized meat export trade. Ice and Refrig. 1, 88–89.

ANON. 1891B. Freezing salmon. Ice and Refrig. 1, 333.

ANON. 1891C. Holding eggs in cold storage. Ice and Refrig. 1, 349.

ANON. 1946. Frozen food industries—Plant layout, cost of processing, marketing. Univ. Arkansas Bur. Res. Inform. Ser. 1, Rev.

ANON. 1954A. The Birdseye story. Quick Frozen Foods 17, No. 2, 55–61, 63, 65, 69–70, 73, 75.

ANON. 1954B. An industry attains maturity. Quick Frozen Foods 17, No. 2, 77, 79, 81, 83, 85, 87, 89, 91, 93, 95, 98–99, 101–103, 105–112.

ANON. 1955A. Frozen vegetables priced under canned in survey. Quick Frozen Foods 17, No. 9, 88–89.

ANON. 1955B. Convenience and quality factors cause housewives to buy more frozen juices. Western Canner and Packer 47, No. 4, 39–40.

ANON. 1955C. Cold cash. Arthur D. Little, Inc., Ind. Bull. 323.

ANON. 1955D. Trends in our eating habits. U.S. Dept. Agr., Marketing Serv. Natl. Food Situation. NFS. 73.

ANON. 1966A. Frozen sea food value soared 26.8% to within sight of a billion dollars. Quick Frozen Foods 29, No. 3, 125–128.

ANON. 1966B. Warehousing and transportation report. Quick Frozen Foods 29, No. 3, 151.

ANON. 1966C. 79 refrigerated warehouses built in two-year period 1963–1965. Quick Frozen Foods 29, No. 4, 99–101.

ANON. 1966D. Freezer storage space capacity USA. Quick Frozen Foods 29, No. 4, 100.

ANON. 1966E. Gross refrigerated storage capacity. Quick Frozen Foods 29, No. 4, 99.

ANON. 1966–1967A. Refrigerated warehouses, 19th Ann. Edition, Sect. 6. Directory of Frozen Food Processors, Quick Frozen Foods, E. W. Williams Publications, New York.

ANON. 1966–1967A. Processors and products—fruits, juice concentrates, vegetables, fish, meat, poultry, prepared foods. 19th Ann. Edition, Sect. 1. Directory of Frozen Food Processors, Quick Frozen Foods, E. W. Williams Publications, New York.

ASHRAE. 1967. Guide and Data Book. Am. Soc., Heating, Refrig., Air-cond. Engrs., New York.

BANKS, M. R. 1954. Capacity of refrigerated warehouses in the United States (as of Oct. 1, 1953). U.S. Dept. Agr., Agr. Marketing Serv., Stat. Bull. 148.

BEEMER, D. B. 1891. Practical cold storage. The handling of poultry. The best way to freeze and pack dressed poultry, etc. Ice and Refrig. *1*, 279–280.

BIRDSEYE, C. 1953. Looking backward at frozen foods. Refrig. Eng. *61*, 1182–1183, 1250, 1252, 1254.

BITTING, H. W. 1955. FF use in preserves, pies, and ice cream. Quick Frozen Foods *17*, No. 10, 129–131.

CARLTON, H. 1946. The freezer locker plant is going commercial. Food Inds. *18*, 1542–1544, 1672, 1674, 1676.

DAVIS, L. L., and RODGERS, P. D. 1948. Planning a frozen food business. Virginia Agr. Expt. Sta. Bull *419*.

DIEHL, H. C., and HAVIGHORST, C. R. 1945. Progress and prospects of the frozen food industry. Food Inds. *17*, 261–278.

EASTWOOD, R. A., and SCABLON, J. J. 1952. Operating cost of 15 cooperative poultry dressing plants. U.S. Dept. Agr., Farm Credit Admin. Bull. *70*.

ENOCHIAN, R. V. 1968. The rise, present importance and future of frozen fresh foods. *In* The Freezing Preservation of Foods, 4th Edition, Vol. 3, D. K. Tressler, W. B. VanArsdel, and M. J. Copley (Editors). Avi Publishing Co., Westport, Conn.

FITZGERALD, G. A. 1956–1957. Fruits and vegetables. *In* Air Conditioning and Refrigeration Data Book, 6th Edition. Am. Soc. Refrig. Engr., New York.

FOX, J. 1955. Spectacular rise of the concentrates emphasized by their tenth anniversary. 1955 Frozen Food Factbook. 39. Natl. Wholesale Frozen Food Distributors Assoc., New York.

FRANKLIN, H. L., and MARTIN, S. 1967. FF per capita consumption rose to 59.389 pounds in 1965. Quick Frozen Foods, *29*, No. 7, 37–43, 154–155.

HARDENBESCH, W. 1955. Four factors shaping frozen meat future. Quick Frozen Foods *18*, No. 5, 126–127.

HASTINGS, W. H., and BUTLER, C. 1956–1957. Fishery products. *In* Air Conditioning, Refrigerating Data Book. Am. Soc. Refrig. Engr., New York.

HAVIGHORST, C. R. 1944. What's Ahead for frozen foods. Food Inds. *16*, 435–439.

HAVIGHORST, C. R. 1955. So you are going into freezing. Food Inds. *17*, 1471–1475.

HAVIGHORST, C. R., and DIEHL, H. C. 1947. Frozen food report No. 2, part 1, transportation, warehousing, frozen food distributional-facilities map. Food Inds. *19*, 3–26; Part 2, marketing. Food Inds. *19*, 149–160.

KAUFMAN, V. F. 1951. Costs and methods of pie stock apples. Food Eng. *23*, No. 12, 97–105.

MANN, L. B., and WILKINS, P. C. 1953. Merchandising commercial frozen foods by locker plants, 1952. U.S. Dept. Agr., Farm Credit Admin. Misc. Rept. *175*.

MAYHEW, W. E. 1952. Cost accounting for the frozen food packer. Quick Frozen Foods *14*, No. 8, 95–98, 326, 328–329.

PAULUS, R. C. 1955. Northwest berry industry trends. Western Canner and Packer *47*, No. 6, 75–79.

PENNINGTON, M. E. 1941. Fifty years of refrigeration in our industry. U.S. Egg and Poultry Mag. *47*, 554–556, 566–567, 570–571.

REDIT, W. H., and HAMER, A. A. 1961. Protection of rail shipments of fruits and vegetables. US Dept. Agr., Agr. Handbook *195*.

SHEAR, S. W. 1955. Trends in United States fruit production and utilization. Western Canner and Packer *47*, No. 6, 70–73.

STAPH, H. E. 1949. Specific heat of foodstuffs. Refrig. Eng. *87*, 767–771.

STEIN, A. 1945. Rise of the frozen food industry. Conference Board Industry Record *4*, No. 10, 1–6.

STEVENS, A. E. 1955. Outlet for concentrates. Quick Frozen Foods *17*, No. 9, 139–140, 204, 206.

STEVENSON, C. H. 1899. The preservation of fishery products for food. U.S. Fish Comm. Rept. *1898*.

TEWSBURY, R. B. 1953. Rail transportation of perishable foodstuffs. Refrig. Eng. *61*, 52–54, 108.

TRESSLER, D. K., VANARSDEL, W. B., and COPLEY, M. J. 1968. The Freezing Preservation of Foods, 4th Edition, 4 vols. Avi Publishing Co., Westport, Conn.

TRIGGS, C. W. 1955. Trends in production, processing and distribution, Fishing Gaz. *72*, No. 6, 83–84.

VANDERVAART, S. S. 1916. Commercial uses of refrigerating machinery. Ice and Refrig. *61*, 179–186.

WENZEL, F. W., MOORE, E. E., and ATKINS, C. D. 1952. Factors affecting the cost of frozen orange concentrate. Quick Frozen Foods *14*, No. 8, 101–102, 244, 246.

WHITMAN, J. M. 1957. Freezing points of fruits, vegetables and florist stocks. Marketing Res. Rept. *196*, US Dept. Agr., Washington, D.C.

WILLIAMS, E. W. 1955A. In which direction is the industry going? Quick Frozen Foods *17*, No. 8, 97–102.

WILLIAMS, E. W. 1955B. Should a wholesaler go into frozen foods? Quick Frozen Foods *17*, No. 10, 49–51, 171–172.

WOOLRICH, W. R. *et al.* 1933. The latent heat of foodstuffs. Tennessee Engr. Expt. Sta. Bull. *11*, Knoxville, Tenn.

CHAPTER 18

Meat, Poultry and Fish Cold Storage and Freezer Storage Rooms

INTRODUCTION

The usual public cold storage warehouse does not, as a rule, contain cooler space devoted to meats, poultry or fish since cooler storage life of these products is of short duration and does not lend itself well to public warehousing. Quite often, however, space in public warehouses may be leased to individuals or companies for their various uses in processing and handling such products. Cooler space is most frequently leased although freezer space may be required in some instances.

MEAT STORAGE

The fresh meat processor usually requires a meat storage room (cooler) with meat rails for hanging carcass meat and a processing room for cutting and processing meat products. Processing equipment is normally furnished by the lessee. The storage and process rooms are sometimes combined into a single area or may be separate rooms depending on the size of the operation and the type of processing setup desired. Several small processors are sometimes accommodated in a single large room partitioned into small areas and each area leased to a different meat processor. Areas may range from a few hundred to several thousand square feet.

When remodeling or building new rooms for lease to any meat processor, local and federal regulations applicable to such areas should be carefully checked with regard to room requirements. These regulations will be found to be strict in the matter of sanitation and other related requirements. Extra drains and sewer lines are often necessary as well as provision for an abundant supply of hot water. Special wall and floor finishes may also be found to be necessary to comply with various codes. Much trouble can be avoided if the proper authorities are consulted regarding applicable regulations before a space is converted or constructed for processing.

Design Items

A typical operation will require a meat storage room with meat rails for hanging meat to be processed. A single rail will usually extend out of the room through a track door and across the loading dock for the receiving of meat from incoming transportation. When extending the rail over the dock area, it is necessary to provide a hinged section or some device for raising the rail so that normal dock traffic may proceed when the rail is not in use. The meat storage rail system within the room should be arranged to suit the requirements of the tenant to give him adequate length to store his anticipated storage load. In addition to the rails, there may be shelving for the storage of processed meat and clear storage space for pallets or push trucks as required. Processing of the meat will usually take place adjacent to the storage area, sometimes in a separate room and sometimes in a low partitioned space taken from the storage space. The processing area will contain the processing equipment required for the particular operation such as saws, grinders and other equipment. Ample electric power is necessary in the processing area to supply adequately all of the processing equipment. In larger operations, an office will be required. This can be set up inside the cooler if no other space is available. These offices are usually small. When set up in a cooler area, they should be well insulated and heated. Electric heat will be found to be the most practical method of heating in most instances.

Special Items of Consideration

Small processors will usually require a much simpler type of space usually consisting of a single room at cooler temperature in which processing is done on quarters which are hand trucked into the room. Shelving may be constructed in the room to hold completed orders. Processing equipment is installed to meet the requirements of the individual operator. As regulations become more strict, the number of small processors tend to become fewer and the larger operations will be found to dominate.

Refrigeration for processing and meat storage rooms sometimes utilizes the existing compressors and system. The cooler facilities of the refrigerated area are converted for meat storage using the installed refrigeration. At other times, refrigeration is installed specifically for the meat processor. These load requirements are not heavy, as a rule, since the meat to be processed is usually received chilled. High humidity is desirable to prevent excess "shrink" of the stored meats and a temperature close to 32°F is required. In an existing installation, the high humidity desired sometimes requires the addition of more evaporator surface with proper controls to bring the humidity to a higher point than is normally carried in a standard cooler. This also limits the use of the remaining space in a cooler, if it is not all used for processing and storage, to cooler products that are not adversely affected by the high humidity. Blower coil units will usually be found in this type of room although brine spray units may be used occasionally and pipe coils will also be found in some older rooms. Some type of defrost, usually automatic, is necessary since these rooms are carried near the freezing point.

Cooking Facilities.—Special care should be exercised with any processor utilizing any cooking processes. Cooling requirements of meat products from elevated temperatures to complete cooling of the product are stringent. This usually requires additional refrigeration of relatively high capacity and should be taken in to account in any lease agreement.

In the processing room will be found various electric saws, grinders and other machinery required in the processing of meat from the carcass to individual steaks, chops, roasts, etc. Also required is a source of hot water in quantity and sometimes steam may be required for processing and cleanup purposes. This room will also be cooled although temperatures are not as critical as in the storage room area. In some instances this area will be cooled by the same refrigeration blowers used for the storage area and separated by a partial partition extending to about rail height. This usually occurs where the processing and storage room are part of an existing installation. Precautions should be taken to limit the load on the refrigeration units in the processing room during the clean up period. Hot water is used extensively for this clean up and a very high instant load can be developed in the refrigeration units which can cause trouble unless it is anticipated and proper precautions taken.

Retail Meat Refrigeration Storage.—Another type of meat operation is that of a wholesaler or processor storing processed meats only for distribution to retail outlets. This requires only a standard cooler storage space since meats of this type are usually well-packaged and do not require as careful control of humidity as carcass meat and exposed cuts. Quite often, a distributing setup will also require a small freezer for the storage of frozen products along with the cooler products stored in the cooler. A small freezer with automatic coil defrost is usually leased for this purpose. This freezer can be built inside of a regular large cooler room with a refrigeration coil of its own or can be a partitioned space in a larger warehouse freezer.

Leased space should be adjacent to or within easy access to dock space with adequate parking space for the lessees trucks and for loading and other requirements necessary to the operation. Most of the processors or wholesalers will use refrigerated trucks with stored refrigerant type plates. In the case of self contained condensing units used to charge the plates, it will be necessary to have plug-in cords available so that the trucks may be refrigerating while at the dock. In the case of the eutectic cold plates with no truck condensing unit, connections to the warehouse refrigeration plant are necessary so that the cold plates may be charged while the truck is parked at the dock. This usually takes place over night.

POULTRY PRODUCTS

Poultry products are usually handled frozen in public warehousing. The freezing process is normally accomplished in the poultry processing plant and the finished product shipped to the warehouse for storage in a frozen state. In some instances, the product may be frozen in the public warehouse freezing facilities and stored. Very little if any cooler space is used in the public warehouse for poultry products.

SEAFOODS

Fish, and more especially shrimp, are held in coolers usually iced down in preparation for processing and freezing. Warehouses having shrimp freezing facilities will also have cooler space available to carry green shrimp until they can be processed. High humidities should prevail in the cooler areas and usually do without too much effort due to the fact that shrimp usually come into the room packed in ice. Other sea products may also be held in cooler storage preparatory to being frozen. This type of warehousing will most often be found in coastal cities, although some processing and freezing may be done inland during peak periods when coastal facilities are short.

FREEZER STORAGE OF MEAT, POULTRY AND FISH

Meat, poultry and fish products, when frozen, can be held in general freezer storage at 0°F or lower. No special requirements must be observed. Except in certain seaports, present day products are usually well-packaged for long freezer storage life.

Many warehouses have freezing facilities that can be used extensively for freezing meat products. Most meat products are not damaged by slow freezing, and when carcass meat is frozen, no warehouse freezer temperature can effect quick freezing on account of the thickness of the carcass which prevents the rapid removal of heat. Many systems are used to freeze halves or quarters of beef. Often they are frozen in a freezer storage room equipped with some extra evaporator surface and auxiliary fans to insure good air circulation. The halves or quarters are hung on racks in the room and allowed to freeze over a period of time of 60 to 72 hr. Some specifications call for complete freezing of the carcass within 72 hr. This is not too difficult to accomplish in a standard freezer room with proper auxiliary fans and good air circulation. If much of this type of freezing is done, it is well to designate a freezer room for this with extra surface evaporator and additional air circulating fans. The room should not be used for storage since the additional air circulation required for freezing and also possible temperature fluctuations as loads are added tend to make an unsatisfactory storage room.

Courtesy of Alford Refrigerated Warehouses, Dallas

FIG. 18.1. BEEF HALVES HUNG ON RACKS IN FREEZER ROOM FOR FREEZING

AIR BLAST FREEZERS FOR MEAT AND FISH

Plant blast freezers are often used for the freezing of poultry and fish products. In freezing poultry, and particularly turkeys, it is necessary to achieve a fairly rapid freeze. At least the surface of the bird must be quickly frozen to set the bloom or color which should be white or faintly pink. Extreme low temperatures are not essential to obtain this bloom but good air circulation is very necessary.

Freezing Shrimp

Shrimp freezing is a rather specialized operation and is usually handled by the shrimp packer. The warehouse usually leases him the space and quite often will sell the packer refrigeration in the form of metered refrigerant for use in the packer's own freezing equipment. Refrigeration is also sold on the basis of the tonnage of frozen product. The freezing may be done in plate type freezers, blast freezers or IQF freezers where the individual shrimp are placed on a conveyor belt and frozen by air blast. Most shrimp are packaged commercially in 5-lb boxes.

Courtesy of Alford Refrigerated Warehouses, Corpus Christi

FIG. 18.2. PACKAGED SHRIMP PRIOR TO FREEZING

Glazing of Shrimp.—After freezing boxed shrimp, the box is opened by raising the lid and the contents glazed with a spray of water. This glazing effectively prevents dehydration of the shrimp and keeps them in good salable condition. When furnishing refrigeration for shrimp freezing and storage, consideration should always be given to the extra load imposed by the glazing process. The refrigeration required to freeze shrimp is greater per ton of product than for most products, due to the glazing process. The glazing is usually done after the shrimp are in storage. A pallet load or so is moved into a work space area, glazing them and then placing them back in the freezer. If this load is not taken into account and proper refrigeration added in the storage room to refreeze the portion melted by the water spray plus some thawing of the product while out of the freezer room, rather serious refrigeration shortage may develop with resultant erratic temperatures in the storage room.

Glazing of Hams.—Green hams intended for long time freezer storage are also glazed, at times, after freezing. This can be done individually by passing the frozen hams through a water bath and then piling them in the freezer storage room; or by direct water spray on a pile of hams in the freezer room. This glazing load is not as high as for shrimp since the amount of water frozen is not great compared to the total product weight.

If a warehouse is in an area containing several packing plants, it is possible that meat freezing can be an appreciable part of the load. An

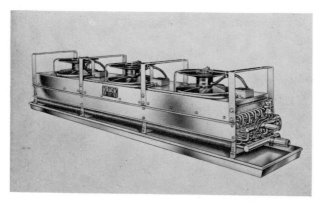

Courtesy of Krack Corp.

Fig. 18.3. Meat Cooler Unit Which May Be Hung Between Meat Rails or in Open Room

acceptable job of freezing on most meat products can be done in a standard freezer with extra cooling surface to accommodate the extra load imposed by the freezing. Temperatures are not too critical and good air circulation is more important than extreme low temperatures. If warehouse space is critical, meats can, of course, be frozen in a blast or tunnel freezer. Bear in mind however, that on carcass meats, no freezing process will allow quick freezing due to the thickness of the carcass. Good air circulation, again, is more essential than extreme low temperatures.

In drawing up a lease for the rental of cooler or freezer space, care should be exercised to spell out exactly what is to be furnished by both parties and also what alterations are to be made, if any, in the space to fit it for the use intended. As a rule, the lessee will furnish processing equipment such as saws, grinders etc. while the lessor will make alterations to the space and put in partitions, trackage, hot water supply and other like items. Alterations are usually figured at cost and this cost added to the price of the lease over the life of the lease.

A good lease must give the lessee what he needs for the satisfactory pursuit of his business, and at the same time, afford the lessor an adequate return on his investment and operating costs. If this is accomplished in the lease agreement, a mutually satisfactory relationship will usually follow.

BIBLIOGRAPHY

ASHRAE. 1967. ASHRAE Guide and Data Book, Systems and Equipment. Am. Soc. Heating, Refrig., Air Cond. Engrs., New York.

HICKS, E. W. 1935. The evaporation of water from meat. Rept. Food Investigations Board, Dept. Sci. and Ind. Res., London.

MADIERA, R. L. 1958. Operational Manual, Vol. I and II. National Institute of Locker and Freezer Provisioners, Elizabethtown, Pa.

TRESSLER, D. K. et al. 1968. The Freezing Preservation of Foods, 4th Edition, 4 vols. Avi Publishing Co., Westport, Conn.

U.S. DEPT. AGR. 1952. Regulations Governing the Meat Inspection. Bur. Animal Ind. U.S. Dept. Agr., Washington, D.C.

WILKING, P. C. 1961. Frozen food locker and freezer provisioning plants in the United States. U.S. Dept. Agr. Frozen Food Locker Branch. Farmer Cooperative Serv., Washington, D.C.

CHAPTER 19

Pear, Apple, and Grape Storage Rooms

PHYSICAL AND BIOCHEMICAL CHARACTERISTICS OF LONG TERM STORAGE APPLES

The growth and physical nature of apples vary widely. All apple varieties adapt themselves to the cooler climates. Whether this particular climatic condition is the result of geographical location in a temperate zone or if the climate acceptability is made possible by the selection of an orchard high in the mountainous elevations of even a semitropical zone.

In the regions below 1,000 ft elevation south of the 35th parallel only a few productive orchards are existent. However, in the 3,000 ft semitropical zone south of the 30th parallel successful commercial orchards are common. Season wise, especially in the river-plains north of the 40th parallel in the United States, varieties of apples may be classified under such names as: June, Summer, Fall, Snow and Winter apples. It is the latter varieties that the cold storage operator receives in vast tonnage for long term storage fruit. The important storage varieties are: Winesap, Rhode Island Greening, Baldwin, Cortland, McIntosh, Rome Beauty, Delicious, Yellow Newtown, and Jonathan.

On the trees, irrespective of variety, the fruit receives a continuous supply of food materials, part of which permits its growth and which goes in to form the fruit. Both before and after harvesting, the fruit goes through a breathing process affected by the temperature, food and moisture available. During all this period, both before and after harvest, the fruit respires taking on oxygen and giving off carbon dioxide similar to the cycles prevalent in animal life. When this respiration stops the apple starts to decay and is no longer edible. Such change can be arrested if the fruit is controlled in an atmosphere held at a temperature of only 30°F. Most apples will keep much longer at 30°F than at 32°F. 30°F is somewhat above freezing temperature of most varieties. The average freezing temperature of apples of all storage house grades is 28.5° F.

A few of the varieties of apples develop internal browning in storage, especially when stored at temperatures below 36°F. Some apple varieties such as Rhode Island Greening, Yellow Newtown and McIntosh are often stored at 36°F to prevent this browning effect. Most other varieties, however, can be held successfully at 30° to 32°F if the fruit is harvested immediately upon maturity. Since all storage fruit continues to respire after harvest there will be some loss of weight in storage unless the relative humidity is held from 80 to 90%. Some packers even prefer, to prevent moisture loss, storing in an atmosphere of 95% RH. When this is done it is necessary to control the mold growth.

The ventilation of the cold storage rooms is usually accomplished by the circulation of air, this primarily to keep the fruit from varying widely in temperature in different parts of the room. The operators must be very careful to maintain a close temperature control in all sections of the room so that it does not vary more than 1°. If apples are kept at 30° to 31°F in a humidity of approximately 90% their storage life can be anticipated to extend as much as 8 months for Red Delicious and Winesap apples and up to 6 months for the Baldwin, Arkansas Black, Rome Beauty and Yellow Newtown varieties.

CONTROLLED COOLING ROOM GASES FOR LONG TERM STORAGE OF APPLES

The "controlled" cooling room storage for apples consists of a gas-tight room refrigerated to the desired cooling room temperature. Provisions must be made for controlling the oxygen, nitrogen and carbon dioxide content in relation to each other.

The initial controlled apple storage rooms were built under the supervision of the Department of Scientific and Industrial Research of Great Britain in County Kent more than two decades ago. Since in ambient air the oxygen level is normally about 21%, the carbon dioxide fraction 0.30% and inert nitrogen nearly 79%, there is a related problem of respiration of the fruit as the live apples continue to respire even at 32° to 35°F and consequently give off carbon dioxide. In a sealed controlled atmosphere room under storage the apples will reduce the oxygen content and the carbon dioxide content will build up. The desired balance is an average of 2.5% to 4% of oxygen and 3% to 5% of carbon dioxide with the remainder nitrogen. Air is added if needed, to prevent a drop in oxygen below 2.5% and quantities of carbon dioxide above 5% are absorbed chemically in a caustic soda scrubber. The low oxygen level and above normal carbon dioxide level greatly inhibit the ripening of the fruit (considerably more than low temperature alone) so that its rate of softening is markedly retarded.

These gas controlled rooms are most effectively sealed with sheet aluminum or galvanized iron, or a heavy metallic foil that can be welded or soldered as a vapor barrier to leakage. Equally effective is a vapor barrier for the floors and ceiling usually of heavy hardboard that has been mopped in asphalt at the joints then thoroughly coated with a mastic as a final vapor seal. The entrance doors and shipping port openings must likewise be equipped with effective gas seals and the "controlled" room tested for gas tightness prior to actual use. The usual test is to put the room under positive pressure of 1 in. of water for 30 min or longer thus assuring minimum air vapor leakage for a reasonable length of time when the room is properly loaded with apples.

In practice, such controlled temperature cold storage rooms should be filled with apples before sealing for long term storage. Once sealed, a daily check should be made of the oxygen and carbon dioxide content since either one, in excess or in deficiency, may cause injury to the apples.

After removal from storage of this nature the shelf-life is several times longer than for ordinary cold storage.

In the United States this method of increasing storage life possibilities has been effectively promoted by Professor R. M. Smock of Cornell University, Professor F. W. Southwick, and others.

COOLING APPLES AND PEARS IN STORAGE ROOMS

In the Pacific Northwest, apples and pears seldom require special precooling facilities. They are usually received and cooled in the same rooms where they are stored, because less handling is required. This procedure is commonly called "room cooling."

The product to be room cooled may be high piled manually or mechanically at the time it is placed in the room or within a few days after entering the room. The fruit entering the room may be already packed or it may be loose in the containers before going to the packing line. In one method of handling, room cooling approximates true precooling; in that case, warm fruit is brought into the room and cooled before packing to remove most of the field heat and then is removed, packed and returned to cold storage for later sale. In room-cooling operations, it is desirable that the cooler provide economical storage and handling facilities and also as rapid cooling as is possible, and that subsequently the commodity temperature be as uniform as possible and be near the minimum allowable temperature for the commodity.

Any discussion of room cooling as a substitute for precooling must recognize that the process has limitations in cooling speed. First, directing air through packages uniformly on a large scale is difficult, so air velocity past the containers is rarely as great as in specially built precoolers;

second, sustained room air temperatures much lower than the minimum optimum storage temperature of the commodity are not permissible. In certain types of precoolers, controlled airflow can be maintained and used to enhance the heat transfer from the commodity and low air temperatures can be used to increase the heat flow rate. This last measure presumes that the commodity will be removed from the precooler before actually cooling any lower than the minimum allowable temperature.

It should be emphasized that adequate capacity to handle the heat load is an absolute prerequisite for a good room-cooling operation.

Cooling Apples and Pears Techniques

The evaluation of cooling performance in terms of "characteristic cooling time" or "half-cooling time" is important. "Characteristic cooling time," or "half-cooling time," is the time required to cool fruit from its initial temperature halfway to the temperature of the cooling air. US Dept. of Agr. researches indicate that containers affect cooling, by: (A) Allowing or denying air access to the fruit for cooling by convection. The great difference between the cooling time required for packed and unpacked fruit under similar stacking conditions illustrates this point. A second illustration is the slow cooling of unpacked fruit wrapped in polyethylene liners. (B) Varying the distance through which heat must be conducted from the center of the container to the exposed surfaces. (C) Imposing varying amounts of thermal resistance to heat flow through the package itself. (D) Shape characteristics which determine how many exposed surfaces the container has for a given stacking procedure.

VARIATION IN ECONOMICAL STORAGE CONDITIONS FOR APPLES

Much of the advisory information that is broadcast in the United States and Canada presumes atmospheric and climatic conditions of the more Western region where refrigerated storage and standardized storage houses design has become well established.

There are many regions of the United States where the local markets are supplied from orchards located at high elevations where the temperatures that are maintained within the 30° to 32°F levels depend upon atmospheric cooling sometime at the temperature of the modulated night air with daytime supplements; little has been written about such storage facilities but millions of bushels of apples are stored annually from the orchards of Arkansas, North Carolina, Alabama, Tennessee, Virginia, Pennsylvania, and New York without the use of mechanical refrigeration. Usually the temperature control is not as close for such operations as might be desired but by building massive rock storage rooms on the mountainside, then keeping this mass cooled by available atmospheric air, the temperature fly wheel effect of keeping the mass at or below 36°F is quite workable. The temperature level of successful commercial storage that is possible with such controlled atmospheric condition without mechanical refrigeration may fall short of the temperature necessary for some types of apples for a satisfactory keeping atmosphere for the 4 to 6 months commonly available to the operators of mechanical refrigerated storage houses, yet it is commonly used by less sophisticated farm storage houses especially in the Southern Appalachian and the Ozark mountain regions.

APPLE PACKING AND STORAGE HOUSES

Estimated Refrigeration Costs

Refrigeration installation costs are estimated from typical bid prices of newly installed systems in any area. Costs vary as the size of each plant and the refrigeration for the largest plants would cost approximately $600 per ton of refrigeration, and for the smaller plants $700 per ton of refrigeration. The estimated installed cost of refrigerating machinery, controls, piping and defrost systems for the various plants is shown in the following Tables: 19.1, 19.2, and 19.3.

TABLE 19.1

ESTIMATED INSTALLED COST AND COST PER TON OF REFRIGERATION MACHINERY, CONTROLS, PIPING AND DEFROST SYSTEMS FOR THREE REFRIGERATED STORAGE PLANTS FOR APPLES AND PEARS

	Storage Capacity		
Item	50,000 Boxes	100,000 Boxes	200,000 Boxes
	Dollars	Dollars	Dollars
Cost/T.R.	667	633	600
Cost of equipment installed	18,000	33,000	60,600

TABLE 19.2

WINTER HEATING LOADS (BTU PER HOUR) FOR PACKING ROOMS AND OFFICES OF THREE PLANTS OF DIFFERENT SIZE IN DENVER, COLORADO 1970

	Storage Capacity		
Item	50,000 Boxes	100,000 Boxes	200,000 Boxes
	Btu/Hr	Btu/Hr	Btu/Hr
Packing room	690,400	675,600	990,400
Office	42,000	47,180	65,000
Total	732,400	722,780	1,055,490

TABLE 19.3

ESTIMATED CONSTRUCTION COSTS OF THREE APPLE AND PEAR PACKING
AND STORAGE HOUSES AT YAKIMA, WASHINGTON

Item	50,000-Box House Total Cost	Cost per Loose Box	100,000-Box Houses Total Cost	Cost per Loose Box	200,000-Box Houses Total Cost	Cost per Loose Box
	Dollars	Dollars	Dollars	Dollars	Dollars	Dollars
Cold-storage room	46,720	0.90	78,230	0.78	144,070	0.71
Machine room and refrigeration	22,470	0.43	39,720	0.39	72,920	0.36
Packing room and office	56,030	1.08	58,930	0.58	105,720	0.52
Heating, plumbing, and electrical equipment	16,270	0.31	21,620	0.22	27,140	0.14
Covered areas	9,360	0.18	14,200	0.14	27,850	0.14
Lot and size preparations	5,500	0.11	7,000	0.07	17,500	0.09
Total	156,400	3.01	219,700	2.18	395,200	1.96

Source: U.S. Dept. Agr., Washington, D.C.

REFRIGERATION OF GRAPES

Grapes are widely grown in the United States, but of the 3 million tons produced annually, over 90% are grown in California. This state produces grapes of the *Vitis vinifera* species almost exclusively (about 2¾ million tons annually) which accounts for over 98% of the production of this European type. Of this amount, 500,000 to 600,000 tons are utilized as table fruit.

Grapes grow relatively slowly, and should be mature before harvest because they do all of their ripening on the vine. Mature here means that state of physiological development when the fruit appears pleasing to the eye and can be eaten with satisfaction. However, grapes should not be overripe as this predisposes the fruit to two serious post harvest disorders: (1) weakening of the stem attachment in some varieties, such as Thompson seedless, which causes the berries to separate from the pedicel attachment; and (2) progressively greater susceptibility to the invasion of decay organisms. Danger of decay is also enhanced if the fruit has been exposed to rain or excessively damp weather before harvest—conditions favorable for the inception of field infections by *Botrytis cinerea* Pers that result in a higher decay potential in the fruit.

Storage Environment

Recommended storage temperatures for *Vitis vinifera* (European or California type) grapes are 30° to 31°F. The relative humidity should be from 87 to 92%. Although temperatures as low as 28°F have not been injurious to well-matured fruit of some varieties, other varieties of low sugar content have been reported damaged by exposure to 29°F.

Storage plants that specialize in storing grapes should provide uniform air circulation in the rooms. Some plants provide precooling rooms where the grapes are cooled to 36° to 40°F before they are placed in storage. In some plants all of the cooling is done in the storage rooms; however only a few have sufficient air movement to cool the fruit as quickly as desired. Experience has indicated that approximately 4,000 to 6,000 cfm per carload of fruit is required for rooms used for precooling. After the fruit has been precooled the air velocity should be reduced to a rate which will maintain uniform temperatures throughout the room (no more than 10 to 20 fpm in the channels between the lugs). Ventilation is required only to exhaust sulfur dioxide and air following fumigation.

The normal change that takes place in grapes in storage involves chiefly loss of water. The first noticeable effect is drying and browning of stems and pedicels. This effect becomes evident with a loss of only 1 to 2% of the weight of the fruit. When the loss reaches 3 to 5%, the fruit loses its turgidity and softens.

Grapes are particularly vulnerable to the drying effect of the air because of their relatively large surface to volume ratio, especially that of the stems. Stem condition is very important as a quality factor and is an excellent indicator of the past treatment of the fruit. Emphasis should be placed on maintaining stems in a green fresh condition, both from an appearance standpoint and because stems become brittle when dry and are apt to break. The stem of the grape cluster, unlike that of other fruits, is the handle by which the fruit is carried, so if breakage occurs (shatter) the fruit is lost for all practical purposes even though the shattered berries may still be in excellent condition. Therefore, careful attention should be paid to those operations which will minimize moisture loss, including prompt and rapid precooling, and reduced air velocity after the field heat is removed. The color of red and blue varieties gradually turns darker in storage. White varieties, such as Ohanez and Thompson seedless eventually turn brown but this is not due to chilling injury.

The most common cause of loss in grape storage is gray mold rot

(*Botrytis* sp.). Prompt cooling, constant low temperatures of 30° to 31°F, and regular fumigation in storage are the chief aids in controlling it. Rains or foggy weather during harvest are the important contributing factors in starting gray mold infections.

The storage life of grapes is affected in large degree by the attention given to selecting and preparing the fruit. Grapes should be picked at the best maturity for storage, especially Sultanina and Ohanez. Stems and pedicels should be well developed and the fruit should be firm and mature. Soft and "weak" fruit should not be stored. The display lug is a satisfactory package for storage since it can be cooled and fumigated easily. Sawdust packages cannot be fumigated effectively, so that it is necessary to fumigate the grapes before they are packed. South African packs of wrapped bunches in excelsior have proved to be good storage packages. Precooling to 40° to 45°F is advised for grapes that are to be in transit a day or two before reaching storage. Special care should be taken that good transit refrigeration practices are used for grapes shipped to storage so that decay will not get started. Grapes intended for storage, except those in sawdust packages, should be fumigated when shipped. It is not good practice to delay fumigation until the grapes reach a distant plant, for in the picking and packing of grapes many berries are injured sufficiently to permit mold to get started unless the fruit is fumigated promptly.

For *labrusca* (Eastern type) grapes storage temperatures of 31° to 32°F and humidities of 80 to 85% are recommended.

The Eastern varieties are not fumigated with sulfur dioxide because of their susceptibility to injury from it. The storage life of the important commercial varieties at 32°F is shown in Table 19.4.

TABLE 19.4

STORAGE LIFE OF EASTERN US GRAPES

	Weeks		Weeks
Concord	4 to 7	Catawba	5 to 8
Niagara	3 to 6	Worsen	3 to 5
Delaware	4 to 7	Moore	3 to 6

TABLE 19.5

STORAGE LIFE OF CALIFORNIA TABLE GRAPES

	Months
Emperor, Ohanez, *Alphonse lavallee* (Ribier)	3–5
Malaga, Castiza (red Malaga), Cornichon	2–3
Sultanina (Thompson Seedless)	1–2.5
Flame Tokay, Alexandria (Muscat)	1–1.5

TABLE 19.6

COMPARISON OF SOME RESPIRATION HEAT VALUES IN COOLING FROM 80°
TO 32°F OVER THREE DAY PERIODS

Kind of Fruit	Heat of Respiration in Btu per Ton of Fruit	Sensible Heat, Btu
Winesap apples	8,000	80,000
Elberta peaches	13,000	80,000
Bartlett pears	19,000	80,000
Chesapeake strawberries	30,000	82,000
Florida oranges	9,000	81,000

BIBLIOGRAPHY

ALLEN, F. W., and CLAYPOOL, L. L. 1948. Modified atmospheres in relation to the storage life of Bartlett pears. Proc. Am. Soc. Hort. Sci. 52, 192.

ANON. 1957. Grapes and grape culture. Americana Encyclopedia 13, 140.

BROOKS, C., and HALEY, C. P. 1934. Soft scald and soggy breakdown of apples. Agr. Res. 49, 55–59.

CARRICK, D. B. 1930. Some cold storage and freezing studies of the fruit of the vinifera grape. Cornell Expt. Sta. Memo. 131.

FISHER, D. F. 1939. Storage of Delicious apples in artificial atmosphere. Proc. Am. Soc. Hort. Sci. 37, 459–461.

FISHER, D. F. 1942. Handling apples from tree to table. US Dept. Agr. Circ. 659.

GERHARDT, F. 1950. Air purification in apple and pear storage. Refrig. Eng. 58, 145–148, 192–194.

HARDENBURG, R. E., and ANDERSON, R. E. 1961. Polyethylene box liners for storage of Golden Delicious apples. US Dept. Agr. Marketing Res. Rept. 461.

HARVEY, J. M., and PENTZER, W. T. 1960. Market disease of grapes and other small fruits. US Dept. Agr. Marketing Res. Rept. 189.

HERRICK, H. F., JR., and SAINSBURY, G. F. 1964. Apple packing and storage houses layout and design. US Dept. Agr. Marketing Res. Rept. 602.

HUKILL, W. R., and SMITH, E. 1946. Cold storage for apples and pears. US Dept. Agr. Circ. 740.

KIDD, F., and WEST, C. 1952. Soft scald and its control in Delicious apples. Proc. Wash. State Hort. Assoc. 48, 97–100.

KIDD, F., WEST, C., and SIEGELMAN, H. M. 1953. Air purification for fruits. Ice and Refrig. 124, No. 4, 15–19.

LUTZ, J. M. 1938. Factors influencing the quality of American grapes in storage. US Dept. Agr. Tech. Bull. 606.

NELSON, K. E., and GUILLOU, R. 1960. Highly marketable grapes maintained by controlled humidity. ASHRAE J. 2, No. 4, 57–61.

NELSON, K. E., and RICHARDSON, H. B. 1961. Further studies on factors affecting the concentration of sulfur dioxide in fumigation atmospheres for table grapes. Proc. Am. Soc. Hort. Sci. 77, 337.

NELSON, K. E., and TOMLINSON, F. E. 1958. Some factors influencing bleaching and wetness of Emperor and Tokay grapes. Proc. Am. Soc. Hort. Sci. *71*, 190.

PENTZER, W. T. 1931. The cold storage of grapes. Ice and Refrig. *81*, 84.

PENTZER, W. T. 1940. Factors affecting the storage of grapes. Assoc. Refrig. Warehouses Spec. Bull. *1*, No. 4.

PLAGGE, H. H. 1942. Controlled atmosphere storage for Jonathan apples. Refrig. Eng. *43*, 4, 215–220.

SAINSBURY, G. F. 1961. Cooling apples and pears in storage rooms. US Dept. Agr. Marketing Res. Rept. *474*.

SMOCK, R. M. 1940. The storage of apples. Cornell Exten. Bull. *440*.

SMOCK, R. M. 1952. Air purification in fruit storage. Ice and Refrig. *123*, 13–17.

SMOCK, R. M. 1958. Controlled atmosphere storage of apples. Cornell Exten. Bull. 759.

SMOCK, R. M., and YATSU, L. 1960. Removal of carbon dioxide from controlled atmosphere storage with water. Proc. Am. Soc. Hort. Sci. 75, 53.

WRIGHT, R. C., ROSE, D. H., and WHITEMAN, T. M. 1954. Commercial storage of fruits, vegetables, and Florida stocks. US Dept. Agr. Handbook. *66*.

ZAHRADNIK, J. W. 1959. Controlled atmosphere apple storage process and its requirements on refrigeration structures and systems. ASHRAE Trans. 65, 589–596.

CHAPTER 20

The Storage of Some Tropical and Semitropical Fruits and Vegetables

INTRODUCTION

Inherently, fruits and vegetables of tropical and semitropical origin are more susceptible to very low storage temperatures than the produce commonly found that has been grown in the more temperate climates. For generations residents of the Gulf States of the United States and the tropical areas from the Rio Grande to the equator have found that yam sweet potatoes, bananas, papaya, citrus fruit, okra, olives, etc. keep better when stored at temperatures above 40°F. Many householders of the temperate zone still cannot understand why such items as sweet potatoes, bananas and lemons store so bady in their household refrigerators set at a temperature most acceptable for milk, apples, meat and similar foodstuffs.

Southern farmers have been called upon often to explain to a novice vegetable grower, who has migrated from the northern Irish potato fields, why his first crop harvested yams that he has stored in his walk-in cooler held at near 35°F have not cured to his satisfaction.

Table 20.1 gives a rundown on the tropical and hot climate fruits

TABLE 20.1

TROPICAL AND HOT CLIMATE FRUITS AND VEGETABLES SUSCEPTIBILITY
TO LOW TEMPERATURES WHEN PLACED IN COLD STORAGE
AT TEMPERATURES BELOW THOSE RECOMMENDED[1]

Commodity	Approximate Lowest Safe Temperature °F	Character of Injury When Stored Between 32°F and Recommended Temperature
Avocados	45	Internal browning
Bananas, green or ripe	56	Dull color when ripened
Grapefruit	45	Scald, internal browning
Guava	45	Browning
Lemons	55 to 58	Internal discoloration
Limes	45	Pitting
Mangoes	50	Internal discoloration
Okra	40	Discoloration
Olives, fresh	45	Internal browning
Oranges, California	35 to 37	Rind disorders
Papayas	45	Breakdown
Sweet potatoes	60	Decay, internal browning
Tomatoes		
Ripe	50	Breakdown
Mature-green	55	Poor color when ripe

[1] Basic data originally compiled mostly by ASHRAE.

and vegetables susceptibility to low temperatures when placed in cold storage at temperatures below those recommended.

Within this chapter the special considerations that should be given to each of these fruits and vegetables are recorded for the information of cold storage managers and operators who may be receiving quantities of these products.

CITRUS FRUIT

Marketing studies show that losses in the distribution of citrus fruit are due to rind breakdown, decay, freeze damage, and overripeness of fruit. In addition, losses may also be caused by changes in the fruit itself. Special attention is given to better methods in the maintenance of quality of fresh citrus fruits, since loss from spoilage continues to be the chief marketing problem. Two reports point up the seriousness of decay and waste. One report states that "out of every 5 acres growing perishable citrus products, the produce of 1 acre is lost through waste and spoilage in marketing channels," and the other report estimates losses "between 10 and 30% the world over from rodents, insects, and microorganisms." The highest losses are reported in the under-developed nations badly in need of food.

Maturity and Quality

The degree of ripeness of citrus fruits at the time of harvest is the most important of the factors determining eating quality. Oranges, like grapefruit, do not improve in palatability after harvest. They contain practically no starch, do not undergo marked change in composition after being picked from the tree, as do apples, pears or bananas, and owe their sweetness to natural sugars contained when they are picked.

Quality is often associated with appearance, firmness, freedom from blemishes, thickness, texture, and color of rind. Actually, the determination of quality should be based on the texture of flesh, juiciness, content of total solids (principally sugars), total acid, aromatic constituents, vitamin and mineral content. The age of the fruit is also important, because immature fruit is usually coarse, very acid or tart, and the internal texture is ricey or coarse. Overripe fruit held on the tree too long may become insipid, develop off-flavors and possess short transit, storage, and shelf life. The importance of having good quality fruit at harvest cannot be over emphasized. The main objective thereafter is to maintain the quality and freshness.

Harvesting and Packing-House Operations

Handling.—In the packing-house, citrus fruits are prepared for shipment by a number of carefully directed operations which are varied to suit the variety or quality of the fruit and the market requirements. The general aim is to keep physiological breakdown and decay to the minimum and to prepare the fruit in an attractive manner.

Citrus fruit storage at the consuming and fruit express distribution centers is much alike for oranges, grapefruit and lemons. The storage atmosphere and temperature for lemons and grapefruit are very similar, each requiring a preferred holding temperature of 58° to 60°F and a humidity of 86 to 88% RH.

The removal of the vital heat of respiration is required for each type of fruit and separate tables of this heat of respiration are given herewith. Adequate ventilation will remove this respiration heat and carbon dioxide.

Storage of Citrus Fruit

The history of the storage of citrus fruit is one of educating the peoples of the colder climates how to store tropical fruits and vegetables. In general, tropical fruits do not respond to long term storage at low temperatures like those fruits and vegetables grown under a more temperate climate. The use of grapefruit as human food does not date back before the twentieth century. Over the first 25 yr of cold storage experience with grapefruit, the cold storage results were often most disappointing when extended for a storage period of more than a few days. Modern practice has made it possible to hold grapefruit in cold storage for 4 to 6 weeks. In California, where the climatic conditions permit the harvesting of grapefruit from the tree over several months, the crop is retained in holding rooms for only short periods of time. The trees serve as the ventilated storage space for some months. In Texas, where the control of the fruitfly demands that all fruit be removed from the tree, usually by April, then the storage period is often extended to 6 to 8 weeks and this preferably at 32°F.

Since grapefruit is a tropical product, it shows much better adaptation to total period of storage if kept at 50°F for a short time before being transferred to the 32°F rooms. By this procedure it is possible to hold much of the harvested grapefruit in the 32°F rooms up to 8 weeks. Some growers adopt a standard practice of holding their crop in rooms held at 50°F 1 week, then transferring it to 32°F colder storage at 86 to 88% humidity for the remainder of the 6 to 8 weeks. Proper stacking of the fruit containers in storage rooms is important to secure uniform

TABLE 20.2

HEAT OF RESPIRATION OF CITRUS FRUITS AT DIFFERENT TEMPERATURES

Temp °F		Btu per Ton of Fruit per Day				
		Oranges California			Grapefruit	Lemons California Eureka
	Florida	W. Navels	Valen- cias	Florida	Cali- fornia Marsh	
32	700	900	400	500	500	700
40	1400	1400	1000	1100	800	1100
50	2700	3000	2600	1500	2000	2500
60	4600	5000	2800	2800	2600	3500
70	6600	6000	3900	3500	3900	5000
80	7800	8000	4600	4200	4800	5700

air circulation and temperature control. The stacks should be at least 2 in. apart and the rows 4 in.; trucking aisles at least 6 ft wide should be provided at intervals.

ORANGES

The orange is the most important of the citrus family. Its exact origin is lost in antiquity but it is generally accepted that the sweet orange originated in Southeast Asia then migrated westward into Europe and North and South America. About the sixteenth century, the Spaniards brought them into Florida and Florida still remains the principal American production center of the sweet orange although for many years California was a very close competitor. Each of these States often exceeds the equivalent of sixty million boxes of fruit annually. Texas, Arizona, Louisiana, Mississippi, Alabama and Georgia produce smaller quantities for commercial production.

The sour orange has never been of importance in North America. In both the European and African sides of the Mediterranean, sour oranges are grown in great quantities and are processed all over Europe in the production of marmalades and similar jam-like table sweets. There are many varieties of the so-called Mandarin oranges. These are usually smaller and most of them are loose-skinned such as tangerines and tangeroes, and similar hybrids of the sweet orange. Since virtually all orange varieties blossom in the spring, growing regions are limited to areas that do not experience night temperatures below 26°F, especially during the bloom season. While North America produces a larger sweet orange crop than any other part of the world, other countries on the American continent that produce abundantly are: Argentina, Mexico, Brazil, Paraguay and Venezuela. In Asia, commercial production is available from Israel and Turkey of the Near East and Japan, China and India. Egypt, Union of South Africa, Algeria and Morocco are commercial producers in Africa, while in Europe, Spain, Italy and Greece furnish much of Europe's consumption.

Prior to 1945, boxed oranges were the principal orange product for the American market. Since then canned and frozen orange juice has absorbed a large portion of the crop surpluses. The 1967 data indicates that a large portion of the production of both Florida and California has been made into juice and frozen.

Florida- and Texas-grown Valencia oranges can be stored successfully for 8 to 12 weeks at 32° to 34°F and 85 to 90% RH. The same requirements apply to Pope's Summer orange, a late-maturing Valencia type orange.

A temperature range of 40° to 44°F for 4 to 6 weeks is suggested for storage of California oranges. Arizona Valencias harvested in March store best at 48°F, but fruit harvested in June stores best at 38°F.

Florida and Texas oranges are particularly susceptible to stem-end rots. Citrus fruits from all producing areas are subject to blue and green mold rot. These decays develop in the packing-house, in transit, in storage, and in the market, but can be greatly reduced if fruit is properly treated with SOPP, or packed with biphenyl pads. Hot water treatments effectively reduce post-harvest decay of nonethylened Pineapple and Valencia oranges. Proper temperature is also a very effective method of reducing decay. However, once storage fruit is removed to room temperature, decay will develop rapidly. Storage of oranges is often complicated by the fact that prolonged holding at relatively low temperatures may induce the development of physiological rind disorders not ordinarily encountered at room temperature. Aging, pitting and watery breakdown are the most prevalent rind disorders induced by low storage temperatures. Generally, California and Arizona oranges are more susceptible to low temperature rind disorders than Florida oranges.

TABLE 20.3

TYPICAL HEAT PRODUCTION OF ORANGES STORED AT THE LISTED
TEMPERATURES PER TON OF FRUIT

Temp °F	Heat Evolved per Ton of Fruit Btu per 24 Hr
32	900
40	1400
60	5000
80	8000

GRAPEFRUIT

While grapefruit was late in coming on to the markets as an edible citrus fruit, it being virtually unknown before 1750, it now comes on to the American and European markets in great tonnages. Of the world supply, about 90% is produced in the United States, especially in Florida, Texas, California and Arizona, the average yearly value exceeding $50,000,000.

There are pink, white and red flesh varieties. The fresh pink variety claims some premium in most world markets. The annual crop in the United States is often seriously affected by frost in the orchards. This often changes the amount of the fruit that must be carried in storage. Some frozen segments are carried in freezer storage and usually are preferred in salads. Grapefruit canned juice has become most acceptable and the grapefruit juicing industry has been very effective in conserving the world surplus of the annual product.

Historically, grapefruit was introduced into Florida by the Spanish but was not grown commercially there until 1880.

Florida and Texas grapefruit is frequently placed in storage for 4 to 6 weeks without serious loss from decay and rind breakdown. The recommended temperature is 50 to 55°F. A temperature range of 58° to 60°F is recommended for the storage of California and Arizona grapefruit.

Decay and rind breakdown are deterrents to long storage of grapefruit and may develop in fruit during storage or following removal from storage. Pre-storage treatment with fungicides will greatly reduce these problems. Also periodic inspections of the stored fruit should be made in order to terminate the storage at the very first symptoms of development of rind pitting or excessive decay.

Extensive studies have been conducted on simulated and accompanied overseas shipments of Florida grapefruit. Export may require 10 days

TABLE 20.4

AVERAGE HEAT PRODUCTION RATE OF GRAPEFRUIT IN STORAGE
AT DIFFERENT TEMPERATURES

Temp °F	Mg CO_2 per Kg per Hr	Heat Evolved per Ton of Fruit Btu per Hr
40	4.86	1070
50	6.92	1522
60	12.60	2770
70	16.0	3520
80	19.0	4180

to 4 weeks of storage in a refrigerated hold and present problems similar to those encountered in refrigerated storage. These tests revealed that Marsh Seedless and Ruby Red grapefruit picked before January retained appearance best when stored at 60°F. With riper fruit 50° to 55°F is a better storage temperature range for export shipments. Very ripe fruit harvested in April and May, however, develop excessive decay following storage at 50° to 60°F.

LEMONS

The lemon tree is a native of India; it has been introduced into the citrus orchards in almost all of the major tropical and semitropical zones of the world. It is somewhat more sensitive to cold than grapefruit or oranges. The handling of the lemon fruit from the tree is quite different from any of that of the other citrus fruits. It usually ripens in the winter, the fruit is actually cut from the tree long before it ripens while it is still green. Usually the cutting is controlled by an inspector who carries a gage that measures 2¼ in. When the tree fruit passes through this gage it is picked, primarily for the summer market, thus a large portion of the lemon crop is picked during the period of least consumption and must be held in storage until the following summer when consumer demand calls for its distribution.

From the green stage the fruit must be "cured" and held in controlled temperature, and humidity conditions which involve ventilation control and adequate air circulation in the storage facility. It is usually desirable to hold the curing temperature at 58° to 60°F, and preferably never below 50°F. Temperatures above 60°F usually introduce undesirable decay bacteria, thus the storage rooms must be adequately refrigerated. At times heat may be required to maintain a narrow temperature range of 58° to 60°F within the room.

The most successful preliminary handling of the lemon production

TABLE 20.5

AVERAGE HEAT PRODUCTION RATE OF LEMONS IN STORAGE
AT DIFFERENT TEMPERATURES

Temp °F	Mg CO_2 per Kg per Hr	Heat Evolved per Ton of Fruit Btu per 24 Hr
32	2.65	580
40	3.70	810
50	10.50	2310
60	13.5	2970
70	18.6	4090
80	28.2	6200

is usually concentrated in associations rather than by individual grower management. The lemons are picked, stored and marketed under strict Association control, this to assure the market that the produce is of uniform quality throughout the marketing year.

In most areas, lemons are cut from the tree while still green and after washing they are graded by color preparatory to the curing and storage.

Four colors are recognized in the sorting process; they are dark green, light green, silver and yellow. The dark green, as the name implies, is the least ripe and, in order from the dark green to the yellow, the fruit represents different stages of ripeness.

The dark green has the longest keeping property and, after curing, the best juice capacity. For long term storage the yellow has poor keeping properties and usually is not stable for more than 3 to 4 weeks, while the dark green has a normal storage life of 5 to 6 months, especially if treated with a water emulsion before they are put in storage.

For sizing most large lemon Association storage houses grade their crop in 3 or 4 sizes as well as 4 colors, subsequently they send their best grade to the retail market and the lower grades to the juicers and other processors. Whether the fruit is of the best or the lowest quality, most lemons must go as mentioned through the process of curing during which they are held at 58° to 60°F and 86 to 88% RH. Like all fresh fruits, lemons give off carbon dioxide while in storage and this by-product must be removed and fresh air and oxygen replaced by careful ventilation. While the fresh air will vary with the different grades of fruit, good practice provides that 75 cu ft per min of fresh air per 600 storage boxes is the minimum for both storage and shipping of the fruit. Citrus fruit storage at the consuming and fruit express distribution centers is much alike for oranges, grapefruit and lemons. The storage atmosphere and temperature for lemons and grapefruit are very similar, each requiring a preferred holding temperature of 58° to 60°F and 86 to 88% RH.

A large portion of the lemon crop is picked during the period of least consumption and must be stored until consumer demand justifies shipment. It is customary for most of this storage to be done near the producing areas rather than at the consuming centers.

Lemons picked green but intended for immediate marketing, such as most of those grown in the desert portions of Arizona and California, are degreened and cured at 72° to 78°F and 88 to 90% RH. This may take from 6 to 10 days depending on the color when picked and on the nature of the lemons. The thin-skinned Pryor strain of Lisbon lemons turn yellow in about 6 days, whereas the thick-skinned old line Lisbon

TABLE 20.6

TIME OF EXPOSURE (HOURS) AT DIFFERENT TEMPERATURES CAUSING INJURY
OF COMMERCIAL IMPORTANCE

Color of Lemons When Exposed	Temperature of Exposure, °F			
	45	40	35	30
Green	12	26	15	18
Turning	14	18	12	14
Green tip	18	12	12	6
All yellow	12	12	18	12

fruit requires as long as 10 days. Lemon storage rooms must have accurately controlled temperature and relative humidity; the air should be clean and should be circulated uniformly to all parts of the room. Ventilation should be sufficient to remove harmful metabolic products. Air-conditioning equipment is necessary to provide satisfactory storage conditions, as natural atmospheric conditions are not suitable for the necessary length of time.

A uniform storage temperature of 58° to 60°F is important. Fluctuating or low temperatures cause lemons to develop an undesirable high color or bronzing of the rind. Temperatures of 52°F and lower cause a staining or darkening of the membrane dividing the pulp segments and may affect the flavor. Temperatures above 60°F shorten the storage life and are more favorable to the growth of decay-producing organisms. A relative humidity of 86 to 88% is generally considered satisfactory for lemon storage, although a slightly lower humidity may be desirable in some locations. High humidities prevent proper curing of the lemons, encourage mold growth on walls and container, and decay of the fruit, whereas much lower humidities cause excessive shrinkage.

LIMES

The lime belongs to the citrus family and resembles the lemon in many ways. The tree, however, is more prickly and the fruit more acid than any of the commercial varieties of lemon.

The lime originated in Southern Asia in the tropical and warmer countries where its juice has been used for some generations in mixing of cold drinks. In the American continent, wild lime trees grow abundantly in the West Indies, Mexico and Central America. Only in thickets of wild limes does Florida and California attempt to compete with the West Indies or Central American market.

In storage, limes are a typical tropical fruit that should be held above 45°F for the best results. Only a small percentage of the limes grown ever reach storage as a commercial fruit due to its extensive use in the production of citric acid and commercial lime juices where it is used as an extract in the manufacture of gelatin puddings and salads.

BANANAS

Bananas are preserved by freezing for use such as in ice cream flavor bases and the preparation of bakery products, e.g., banana cream pie. Von Loesecke (1950) in his book on bananas discusses a number of varieties. The only variety commercially frozen at present is the Gros Michel. This is the yellow variety ordinarily sold in retail stores and fruit stands in the United States.

Since bananas used in the bakery usually must be mashed before mixing, very ripe bananas are usually selected for bakeshop freezing. To have available a continuous supply of very ripe bananas, freezing is often practiced some weeks prior to the periods of off-season supply, to bridge over the time gap when ripe bananas are not available on the local markets.

The refrigerating equipment and design factors used in the banana industry can best be understood if there is a general knowledge of the background of this tropical fruit. Bananas are available in tropical climates around the world, but the United States imports bananas mainly from Central American countries, Mexico, the West Indies, Colombia and Ecuador. There are many varieties of cultivated bananas, botanically

TABLE 20.7

AVERAGE WEIGHT PER BUNCH (LB) OF BANANAS IMPORTED
INTO THE UNITED STATES FROM JANUARY TO JULY 1945[1]

Country of Origin	1945
Mexico	36.1
Guatemala	61.8
Honduras	60.2
British Honduras	24.9
Nicaragua	33.2
Costa Rica	56.4
Panama	58.0
Canal Zone	59.0
Colombia	35.6
Ecuador	—
Bahamas	—
Cuba	22.2
Haiti	30.2
Dominican Republic	39.9
Jamaica	35.0
French West Indies	31.0
Average	50.0

[1] Official Statistics, US Dept. Comm., Washington, D.C.

TABLE 20.8

VITAMIN CONTENT OF BANANAS AS COMPARED WITH OTHER FRUITS AND VEGETABLES[1]

| Commodity | Vitamin Content (per 100 Gm Fresh Edible Portion) | | | |
	Vitamin A (Carotene), IU	Thiamine μg	Riboflavin μg	Ascorbic Acid Mg
Bananas	50–332	42–62	87–88	3–11
Apples	75	25	73	5–8
Oranges	62–286	78	32–45	52–72.5
Grapefruit	0	—	37–45	27–45
Tangerines	932	84	58	41.9
Pears	10	25	76	3–5
Carrots	2100	72	66	3–5
Cabbage	0–100	81	44	50–90
Tomatoes	1000	78	52	21–24
Cantaloup	300	60	73	26–34
Peaches	1700	20	64	8

[1] The levels should be taken as comparative and not necessarily as absolute.

related but having various handling characteristics and disease resistance. The principal banana varieties grown commercially are the Gros Michel, Dwarf Cavendish, Giant Cavendish, and the Claret or Red banana. Other than the red bananas, and the so-called finger varieties, these varietal differences are not generally distinguishable to most commercial consumers. Most of the bananas marketed in the United States are either of the Gros Michel or Giant Cavendish variety. Although bananas are susceptible to several diseases, those which seriously affect the economics of the industry are Panama Disease, Sigatoka, and Dry Rot or Moko.

Harvesting and Transportation

Bananas do not ripen satisfactorily on the parent plant in the tropics; therefore, the fruit is harvested green and at a stage of development which will allow it to be transported in refrigerated carriers to the country of destination and arrive there still green. Another reason for

TABLE 20.9

CHEMICAL COMPOSITION OF DIFFERENT VARIETIES OF RIPE BANANAS[1]
EXPRESSED AS PERCENTAGE OF FRESH PULP

Component	Gros Michel	Lady Fingers	Lacatan	Plantain	Red[2] Bananas
Moisture	75.9	70.6	71.6	63.8	73.3
Reducing sugars	10.73	6.19	8.15	18.89	4.10
Nonreducing sugars	6.12	13.38	10.01	0.00	16.08
Starch	2.93	4.13	6.54	11.69	4.12
Total carbohydrates	19.78	23.70	24.70	30.58	24.30
Protein	0.81	1.49	1.04	1.16	0.48
Crude fat	0.47	0.30	0.40	0.30	0.24
Pectin	0.34	0.57	0.41	0.43	0.62

[1] These analyses are for fruit the peel of which had developed a full yellow color.
[2] Unpublished data, J. T. Manion, United Fruit Co., Res. Dept., 1933.

cutting the banana green is that the susceptibility of the fruit to handling damage is greatly increased as ripening advances. However, even
with green fruit, precautions must be taken to minimize damage in
transit and at the various points of transfer. Starting at the plantation in
the tropics, one of the first steps taken after cutting, is the attaching
of a loop of heavy string to one end of the stalk which facilitates hanging from overhead conveyors, cutting stands and in ripening rooms; the
fruit is then transported, usually on padded carts, to receiving stations
where it is washed to remove spray residue, dust, and other foreign
matter that may either cause damage or detract from the appearance
of the fruit. After the stem is washed, a transparent polyethylene sleeve
is placed over the fruit to provide protection against abrasive scarring
during transportation. It has the supplementary advantage of enveloping
the fruit in a high humidity atmosphere resulting in fresher arrivals at
destination.

In order to minimize the handling of this tender fruit as much as
possible, the banana industry is now packing a high percentage of
production in boxes at the tropical plantations, thus eliminating many
of the sources of handling damages. The refrigerating industry should
recognize the difference in handling and ripening tropical boxed fruit
as well as stem fruit.

To prevent stems and boxes of bananas from ripening during shipment, the holds of the refrigerated steamships are precooled so that the
field heat will be removed as rapidly as possible. The fruit is reduced
to a carrying temperature of about 56°F (slightly higher or lower depending on variety), which is maintained throughout the journey to the
domestic port. The stems and boxes of fruit are unloaded at the domestic
wharf with mechanical unloading equipment. The discharge of a banana
cargo must be a continuous and rapid operation from the time the ship's
hatches are first opened until all the fruit is unloaded and stowed in
refrigerator cars or trailer trucks for distribution to food warehouses in
various cities. The bananas are protected as much as possible from
damage by exposure to excessively high or low temperatures. In the
winter, where the climate requires, provision is made to heat the discharging areas so that the fruit will not be exposed very long to temperatures lower than 55°F. The bananas are inspected periodically while
enroute to consignee to assure safe delivery temperatures. Only refrigerated, insulated, fantype railroad cars and refrigerated trailers are recommended for banana service. Refrigeration is required in the summer,
and heat is provided in the winter as necessary.

Bananas are sold by the importer to wholesale jobbers in carload or truckload lots of whole stems, or tropically packed boxes weighing 40 lb net. Bananas should be unloaded as soon as possible after arrival at the ripening plant. Prompt unloading and careful handling will minimize damage. The fruit must be brought under control as soon as possible to assure successful ripening on a predetermined schedule which is possible only in properly equipped rooms.

Ripening

During ripening, enzymatic changes take place in the pulp of the fruit. The pulp of the green banana contains about 20% starch. As ripening progresses, this starch is converted to sugars, principally sucrose, dextrose, and levulose. Color changes in the peel occur simultaneously with the pulp change and are also enzymatic. Rates at which these changes take place depend upon the relative activities of different enzymes, which in turn, are stimulated or depressed by the physical conditions to which the fruit is exposed. Green bananas contain two classes of pigments: chlorophyll (green) and carotene (yellow). During ripening, the chlorophyll is gradually destroyed by enzymatic action allowing the carotene to become more and more evident.

The controlling factors in banana ripening are generally recognized to be temperature, humidity and ventilation.

Temperature.—Ripening characteristics of different lots of bananas vary with the country of origin, botanical variety, season of the year, maturity of the fruit when harvested, and other factors. It is impractical to set down precise ripening temperatures that will cover all of the foregoing variables. Suggested ripening temperatures are for bananas with average characteristics. Approximate temperatures for ripening average shipments of bananas are indicated in Table 20.10. This temperature is approximately correct only if ethylene gas is administered on the first day, and the room is ventilated 24 hr after the application of ethylene. If, for any reason, ethylene is not used, initial temperatures

TABLE 20.10

APPROXIMATE TEMPERATURE FOR RIPENING BOXED BANANAS
ON 5 DAY SCHEDULE

Day	°F
1st	64
2nd	62
3rd	60
4th	58
5th	57

must be 3 to 4° higher than that shown. Correct ethylene dosage is covered under a subsequent heading.

Banana Ripening Rooms.—Banana ripening rooms have construction requirements similar in basic principle to ordinary cold storage vaults. Rooms must have adequate insulation, heating and cooling equipment, humidifying equipment, controls, and be as nearly airtight as modern construction techniques permit.

Room Sizes.—Banana rooms vary widely in their dimensions, depending to a large extent on the warehouse layout. The number and capacity of rooms are dependent upon the individual shipping volume of the wholesaler. It is generally agreed that at least three rooms are necessary if an operator is to expect any degree of success in achieving ripening and shipping schedules. Ideally any individual room should be provided for each days business. Capacities are generally based on carlot or part carlot units with 2 carlot, 1 carlot and ½ carlot sizes most generally used. Again the practice of the individual operator must be considered, as carlot shipping units may range from 500 to slightly over 800 boxes.

Room Types.—Room types fall into three general categories: (1) fork-lift room, (2) combination stem-hanging and boxed-fruit rooms, and (3) boxed-fruit rooms.

Airtightness.—Ripening bananas produce ethylene, volatile esters, and carbon dioxide. In addition, ethylene is applied artificially at the initiation of the ripening cycle to permit use of a safe temperature range while meeting schedules. The concentration used in ripening bananas is 0.1%, or 1 cu ft of ethylene gas per 1,000 cu ft of room air space. Retention of gases is essential if processing cycles are to be valid and predictable, hence the need for airtight rooms. Ethylene is explosive in air at a concentration of approximately 3 to 27%. The dangers of over-application of ethylene have been minimized by the almost industry-wide use of 4 oz, 3 cu ft (lectern-bottle size) cylinders.

Refrigeration.—It is desirable, even with high initial installation costs, for each ripening room to have a completely separate system connected to a single condensing unit. Although installations having standby units are sometimes used instead of independent systems. Unit coolers should be installed within the ripening room. Systems in which cooling coils are installed atop rooms, and the air ducted in, are unsatisfactory due to difficulties in maintaining room tightness.

Heating.—During normal ripening cycles, heat is generally required only during the first 24 hr to raise pulp temperature to the desired level for application of ethylene gas. Capacity should be sufficient to raise the pulp temperature at the rate of 2°F per hour.

Humidity.—Humidity plays a definite and important role in the processing of bananas. A long program of tests was conducted in which temperatures, humidities, time period, and handling conditions were varied. In the case of low humidity conditions, it was found that bananas developed an unseen susceptibility to handling damage. High relative humidity in ripening (90 to 95%) contributes greatly to better appearance, salability and product life of bananas. Tests show that weight loss in boxes ripened in low humidity environment are much higher (up to 2 lb per box) than in boxes ripened under high humidity conditions.

Packing and Shipping

Stems on bananas are removed from the ripening rooms at the completion of the ripening process and transported to the packing area which is usually adjacent to the banana rooms. Ripe fruit is extremely susceptible to bruising injury; therefore, the use of mechanized equipment is highly recommended for transferring ripe stems from the rooms to the packing location. Extreme care should be observed throughout the cutting and packing operation.

AVOCADOS

The avocado is also known as the alligator pear and belongs to the tropical American group. Native especially to Central America and Mexico, of recent years it has been grown successfully in the West Indies, Hawaii, Southern Florida and Southern California.

The fruit is often pear shaped and may be, according to variety, green, red or purple skinned. It has invaded the North American market primarily as a salad made of the fruit while the tropical zones have consumed large quantities of avocado pulp in their diet. The other byproduct of the avocado which is commercially distributed, is oil for salads although some is used in making soap and other cosmetic products.

The tree is relatively small and the fruit often develops black spots which renders it more difficult to market. A select avocado does have a tendency to age to a greying color within the flesh and especially around the seed; sometimes however, this darkening is the result of storing at a low temperature. It is recommended that all avocados be stored at temperatures above 45°F and not over 55°F.

The Fuerte variety is the principal avocado grown in California and is especially suited for the preservation of purée products; however, some of the other varieties, especially those of the Guatemalan and

West Indian plantations types, have a yellowish green color and their own characteristic flavor and can be treated to prevent discoloration in cold storage.

GUAVAS

Many persons think that guava is the best fruit of the tropics. The market for guavas in the United States is practically undeveloped. The Lupi variety contains approximately 500 mg of ascorbic acid (Vitamin C) per 100 gm; the Kauai B strain, because of its small amount of seeds; and a Hawaii University selection B-30, because of its very desirable red color, are being planted in commercial Hawaiian orchards for the express purpose of developing a superior frozen guava purée. A frozen guava purée can be used in the preparation of a number of products including beverages and flavor bases for frozen desserts.

The guava is a delicious fruit that blossoms with fragrant white flowers, the most common being of the bush tree type. The fruit is usually somewhat larger than a goose egg and is pear shaped. The guava is full of seeds, fragrant and sweet. Its principal use is in making jellies and purées, the latter being stored as a frozen product.

In North America, the guava is now grown extensively in Southern Florida and California. No great progress has been made to date in quick freezing guava slices.

MANGOS

The mango tree is tropical in nature and several attempts to introduce them in climates of occasional frosts have been unsuccessful. It is a native of Southeastern Asia where the large trees furnish much of the lumber for housing. The trees grow best on sandy soils that are very well drained; they thrive best where frequent rainfall keeps the ground cool. Under tropical conditions the trees grow to a height of 40 ft. The mango fruit is considered by the natives of Southeastern Asia as the best it has to offer. Each fruit contains one large flat seed. The edible pulp of the fruit is soft.

The American supply comes mostly from the West Indies and Mexico although some are grown in the extreme southern areas of Florida and California. The US Dept. of Agr. has the responsibility of inspecting the mangos as they enter the United States. In the American refrigerated warehouses they should not be stored at temperatures below 50°F, otherwise the flesh will turn brown.

The Hawaiian Agricultural Experiment Station is testing a number of mango varieties and some of them show promise as a frozen product. The Joe E. Welch produces the best frozen product (Sherman 1955). The Hayden and Zill varieties are also being studied.

In Mexico the Manzanilla variety is given preference in many markets.

PAPAYA

The papaw of the tropics is called papaya; it is a small palm-like tree with a stem similar to the palm and without noticeable branches—the leaves are its branches. Between the leaves yellow blossoms appear and in hot climates develop into a cantaloup-like fruit which has many black seeds; some varieties have fruit 15 in. long while others do not grow more than 8 or 9 in. in length. They ripen best in humid climates at temperatures between 78° and 90°F and should be firm ripe before shipment at temperatures between 45° and 48°F. When so processed they will keep for 2 or 3 weeks.

The natives of the tropical regions cook the unripe papaya for a food not unlike summer squash. When ripe the papaya is served as fresh fruit similar to the way residents of the temperate zones serve cantaloup. When cut in small pieces, some papaya is processed and marketed in a quick frozen state. Frozen papaya is usually stored at −10°F. More of the marketed frozen papaya, however, comes to the freezer storage warehouse as purée. Large quantities of papaya is marketed as canned fruit in Number 2½ cans not unlike the canned sweet potatoes of the United States.

Utilization

The freezing preservation of papaya as pieces (Sherman, Duernberger, Seagrave-Smith, and Shaw 1952), and as purée (Seagrave-Smith, and Sherman 1954), has been under investigation at the Hawaiian Agricultural Experiment Station. The Solo variety, most commonly grown for fresh fruit in the Islands is also being used for freezing.

The papaya is used extensively in Southeast Asia as the trees grow wild or are cultivated in strains. The fruit varies in characteristics, both in size and edibility. One of the problems in cultivating papaya is that it is very difficult to keep the strains from cross-pollinating, thus to preserve the quality of any one variety it is generally necessary to have acreages set many miles from any other papaya origin to preserve their identity. The juice is much sought after for its vitamin value and much of that that appears on the American market is in the form of canned papaya juice both frozen and unfrozen.

There are extensive plantations in Hawaii and Mexico; some papaya has been successfully grown in Southern California, Florida and Texas but it is strictly a tropical plant and should be cultivated in areas where the separate strains can be kept well apart in order to preserve their identity and useful characteristics. The papaya should never be stored below 45°F. They do not keep well even at this temperature and in prolonged storage at 45°F they become inedible.

OLIVES

While the olive tree has been cultivated for many centuries, primarily for its oil content, in the regions of the Mediterranean, the olive is utilized very extensively as a dinner food. It is not uncommon to find less prosperous workmen of the Mediterranean region munching brown bread and ripe olives with tea or coffee as their total meal. The oil content of the ripe olive is sufficient to furnish these workmen with their required fats for the days labor. In Italy and some parts of Turkey and Greece, the olive is second only to the grape as a horticultural crop.

Over many of the mountain areas, at low elevations, wild olives grow abundantly.

Historically it is of interest that, prior to the Christian era, olives and olive oil were the luxury of the wealthy, especially in the Grecian, Turkish and Italian regions. Later this use and cultivation of olives spread to Spain and thence to Mexico and other Latin American countries. As a tropical fruit fresh olives should be stored at temperatures of approximately 45°F for best results. When stored at lower temperatures they have a tendency to brown.

OKRA

Okra is an edible vegetable member of the mallow family which includes the flowering hibiscus, some strains of the rose of sharon, and the cotton plant. It is extensively grown in the southern United States and some tropical countries. In its growing habits it resembles the cotton plant. Okra is a vegetable used especially to give body to soups, stews and catsups. It is highly mucilaginous. Okra as harvested and piled up, has a tendency to heat readily from its own vital heat, even to the point of steaming when packed in deep baskets. Thus for immediate storage okra should be ventilated and cooled to about 40°F to prevent discoloration by its own spontaneous heating. Okra has never found an extensive market in the temperate zones except with those who have lived for a period in the southern states of the United States.

TOMATOES

Mature green tomatoes can not be successfully stored at temperatures that greatly delay ripening. Tomatoes held for 2 weeks or longer at 55°F may develop an abnormal amount of decay and fail to reach as intense a red color as tomatoes ripened promptly at 65° to 70°F. Temperatures of 65° to 68°F and 85 to 90% RH are probably used most extensively in commercial ripening of mature green tomatoes. At temperatures above 70°F decay is increased. A temperature range of 57° to 60°F is probably the most desirable for slowing ripening without increasing decay problems. At this temperature the more mature fruit will ripen enough to package for retailing in 7 to 14 days.

Storage temperatures below 50°F are especially harmful to mature green tomatoes; these temperatures make the fruit susceptible to *Alternaria* decay during subsequent ripening. Increased decay during ripening occurs following 6 days exposure to 32°F or 9 days at 40°F.

Firm ripe tomatoes may be held at 45° to 50°F and 85 to 90% RH overnight or over a holiday or weekend. Tomatoes showing 50 to 75% of the surface colored (the usual ripeness when packed for retail) can not be successfully stored for more than 1 week and be expected to have a normal shelf-life during retailing. Such fruits should also be held at 45° to 50°F and 85 to 90% RH. A storage temperature of 50° to 55°F is recommended for pink-red to firm-red greenhouse-grown tomatoes.

When it is necessary to hold firm-ripe tomatoes for the longest possible time with immediate consumption upon removal from storage, such as on board ship or for an overseas military base, they can be held at 32° to 35°F for up to 3 weeks with some loss in quality. Mature-green, turning or pink tomatoes should be ripened before storing at this low temperature.

BIBLIOGRAPHY

ANON. 1948. Foreign Agr. Circ. *FDAP-5–48*. Office Foreign Agr. Relations, US Dept. Agr., Washington, D.C.

ANON. 1950. Fruit Dispatch Co. Circ. *14*, 3rd. Rev.

CALDWELL, N. E. H. 1938. The control of banana rust thrips. Queensland Dept. Agr. and Stock, Div. Plant Ind. Res. Bull. *16*, Australia.

EDDY, W. H. 1933. Nutritive value of the banana. Teachers College, Columbia Univ., New York.

GRIERSON, W., and NEWHALL, W. F. 1953. Degreening condition for Florida citrus. Proc. Fla. State Hort. Soc. *66*, 42–46.

GRIERSON, W., and PATRICK, R. 1956. The sloughing diseases of grapefruit. Proc. Fla. State Hort. Soc. *69*, 140–142.

HARDING, P. L., SOULE, M. J., and SUNDAY, M. B. 1957. Storage studies on Marsh grapefruit, 1955–56 season. I. Effect of nitrogen and potash fertilization on keeping quality. II. Effect of different temperature combinations on keeping quality. US Dept. Agr. *AMS-202.*

HARDING, P. L., and WISANT, J. S. 1949. Terminal market storage of Florida oranges. US Dept. Agr., HT&S Office Rept. *218.*

JOSLIN, C. L. 1936. Bananas. Med. J. *29,* 1007.

MANESS, H., and TUCKER, R. G. 1946. World banana production and trade. US Dept. Agr., Washington, D.C.

MANN, C. W., and COOPER, W. C. 1936. Refrigeration of oranges in transit from California. US Dept. Agr. Bull. *505.*

PHILLIPS, R. V., and GRIERSON, W. 1957. Cost advantage of bulk handling citrus through the packing house. Proc. Florida S. Hort. Soc. *70,* 178. Florida Dept. Agr., Winter Haven, Fla.

POLAND, G. L., and WILSON, R. M. 1933. Banana chilling studies. I. Effect of low temperatures on appearance and edibility. United Fruit Co. Res. Dept. Bull. *46.*

ROSE, D. H., BROOKS, C., BRATLEY, C. O., and WINSTON, J. R. 1943. Diseases of fruits and vegetables, citrus fruit and other subtropical fruits. US Dept. Agr. Misc. Publ. *498.*

ROSE, D. H., COOK, H. T., and REDIT, W. H. 1951. Harvesting, handling and transportation of citrus fruits. US Dept. Agr. Bibliographical Bull. *13.*

SCHOLS, E. W., JOHNSON, H. B., and BUFORD, W. R. 1960. Storage of Texas red grapefruit in modified atmospheres—A progress report. US Dept. Agr. *AMS 4-414.*

SEELIG, R. A., and STANLEY, L. T. 1956. Fruit and vegetable facts and pointers, bananas. United Fruit and Vegetable Assoc., Washington, D.C.

SIMMONDS, N. W. 1959. Bananas. Longmans, Green & Co., New York.

SMITH, R. A., and COOK, A. C. 1962. Bananas, world production and trade. US Dept. Agr., Foreign Agr. Serv. *FAS M-128.*

VON LOESECKE, H. W. 1950. Bananas. Interscience Publishers, New York.

WARDLAW, C. W. 1959. Banana Diseases. Longmans, Green & Co., New York.

WRIGHT, R. C., ROSE, D. H., and WHITEMAN, T. M. 1954. The commercial storage of fruits, vegetables and nursery stocks. US Dept. Agr. Handbook *66.*

WHITEMAN, T. M. 1957. Freezing points of fruits, vegetables and florist stocks. US Dept. Agr. Marketing Res. Rept. *196.*

CHAPTER 21

Tree Nut and Peanut Cold and Freezer Storage

INTRODUCTION

The manager of a cold or freezer storage warehouse who anticipates annual shipments of nuts of varying keeping qualities can expect to find that those most common in American cold storage houses are: native peanuts, pecans, walnuts, almonds, Chinese chestnuts and imported Brazil and cashews. The pre-storage harvesting treatment and carelessness in preparing any of these nut crops for delivery to the warehouses is often as important to the final quality as the cold or freezer temperatures in the storage rooms. The warehouse manager must be fully alert to the quality of the nuts delivered if he is intent on delivering a first class product from his doors.

EDIBLE NUTS IN COLD AND FREEZER STORAGE

Temperatures, relative humidity and air purity are of paramount importance to maintain top quality of edible nuts in long term cold or freezer storage. Palatability and nutritive value can be maintained for many months if these three factors are controlled in proper relation to each other.

All nuts have a high oil content and are subject to rancidity, insect and rodent infestation, odor transmittal, mustiness and weight loss by drying out if not held in a controlled atmosphere at low temperatures. Further wide variations in temperature, humidity and air purity control bring about changes in texture, color, aroma and flavor. Good cold and freezer storage houses practically preclude as storage of nuts any rooms with onions, garlic, cheese, meat, chocolate, fresh fruits and cantaloups. Packaging can have a marked effect on successful nut storage, especially in the keeping of shelled nuts. A highly absorbent package of material will present an oily saturation over the contained nuts part and this may even lead to a rancid odor. It is recommended that all nuts be packaged in nonabsorbent materials.

Well-packaged nuts can be safely stored in refrigerated rooms with candies, canned and bottled goods, rice and most dried fruits. The temperatures and humidities for nut storage vary greatly with the crop to be stored. These will be reviewed separately.

261

Almonds

The almond is one of the oldest edible nuts known to modern man. Genesis 43:11 commands: "Take the best fruits in the land in your vessels, nuts, spices and almonds." Since it is an early blooming tree, it flourishes best in those milder climates well-protected from spring-time arctic winds. The first trees of history were apparently indigenous to Western India, Persia and Eastern Turkey but were extensively grown in Palestine and Egypt by the time of Abraham, Isaac and Jacob (1700 B.C.).

With the migration westward, the almond tree was carried into Southern Europe and thence even northward to England. There is a question whether the Romans were responsible for the introduction of the tree along the Mediterranean into Spain and Portugal north-ward along the English Channel or whether the Arabs introduced it into Spain and Portugal during the Islamic invasion of this region and then subsequently Spain introduced it into England about the sixteenth or seventeenth century. The almond tree is small, not unlike the peach, and both are prized for their attractive blossoms. Several varieties of almonds are cultivated as ornamentals.

The earliest almonds were introduced into the United States by the Spanish missionaries, moving into Texas and California from Mexico. This introduction of almond trees was not successful and the almond production in the United States remained inconsequential until after World War I. The United States depended, until this period, for its almond supply on Spain, Italy, Portugal, France, Germany and Great Britain, although there were some importations from India, Iran and Iraq.

The modern US market is supplied more and more by the California Almond Growers Exchange. By 1962, this Exchange built their own cold storage facilities with the capacity of 4,000 tons of shelled almonds. They moved into the American and foreign markets with 12 grades of the shelled product and an equal number of unshelled packs.

By 1968, the Western United States, especially the Almond Growers Exchange, California, is one of the world's major producers and exporters. Spain, which has been the principal supplier of the United States, is now exporting primarily to Northern Europe and South America. The Mediterranean and Southern Asiatic countries of Iran, Turkey, Greece, Portugal and Italy are likewise increasing their annual exportation to the world markets, primarily outside of North America.

Brazil Nuts

Brazil nuts are imported from South America, primarily Brazil and Bolivia. The Brazil-nut tree is one of the largest in the Amazon Valley. They are found in indigenous forests above the Amazon flood lines. The tree bears fruit in a huskless capsule that is 3 to 8 in. in diam that has a hard shell like covering. Each pod contains 12 to 20 nuts arranged as segments within the pod. Brazil-nut trees prefer a well drained clay soil on the higher levels away from swamps and streams. Brazil nuts do not store well, especially in their Brazilian climate and much of the exporting is done in the original pods to be shelled, classified and packaged preparatory to being placed in cold storage. In a typical year 8,000 to 10,000 tons of in-shell Brazil nuts are imported to the United States from a total exportation averaging 25,000 to 30,000 tons a year from Brazil and Bolivia of in-shell Brazil nuts which is supplemented by approximately 5,000 tons of shelled Brazil nuts that are exported by Brazilian and Bolivian merchants.

Nearly all Brazil nuts are harvested on the ground and picked up by the natives. Since the uncultivated native trees are very numerous and are highly productive, the total annual collection is only a small percentage of the annual available tree crop. The Brazilian traders and commercial collectors export part of these from their own trading posts, then deliver large quantities to the shelling centers at Belem and Manaus, Brazil, and to four shelling plants operating in the Bolivian section of the Amazon Valley. Currently the trend in both Brazil and Bolivia is to do their exporting of shelled rather than in-shell nuts. This furnishes income to a domestic labor force of 6,000 to 10,000 workers of the Brazil nut industry of these two nations, and increases the profit to the Brazilian and Bolivian merchants. Many thousands of trees remain unharvested each year, the annual production depending mostly on the foreign demand. The nuts are not economically stored in the Amazon Valley thus the annual carryover of the crop is nearly nil. In the American and European markets Brazil nuts (some times sold under the caption of "nigger toes", cream nuts or paranuts) are sold in the shell or packaged as shredded, salted, candied, roasted or ground. Woodroof (1967) reports the proximate composition of the Brazil nut in Table 21.1.

The Brazil nut has a short shell-life, and unless kept in appropriate storage conditions will soon show marked decay. For local nut marketing in the United States and Europe, it is usually distributed at the

TABLE 21.1

COMPOSITION OF BRAZIL NUTS

Component	%
Moisture	2
Protein	16.3
Extract oils	68.3
Fiber	6.6
Ash	3.61

Christmas season of the year when the new crop nuts are shipped under seasonal low temperature conditions.

Cashews

The cashew tree is native to the American tropics. It is a low spreading evergreen that produces many products. The principal import to the United States and northern Europe is the roasted cashew nut which comes today mostly from India.

The nut is less than 1 in. long and of kidney-like form and occurs in small pods varying in size from plum to pear. The nut is submerged in a caustic oil within the pod. This oil is toxic and strong enough that the people of India use it as a wood preservative to protect floors and walls against white wood ants. The Indians also use this oil in pharmacy to regulate stomach disorders and for gargles.

The nut is first roasted to drive off the caustic taste and odor and make the nut nontoxic.

Today, the imports of roasted cashew nuts to the United States are rapidly increasing but on account of the nature of the liquid within the cashew pod, the US Public Health Department has had to enter into an agreement with the Cashew Export Promotion Council of India to enforce compulsory pre-shipment inspection and quality control.

Cashews keep well in cold storage. Most shipments of the salable nuts come in vacuum packed cans or bottles and are stored below 45°F. When not vacuum packed, they are placed in hermetically sealed metal cans with carbon dioxide gas replacing all air in the containers. Today, the cashews as a tree nut is second only to almonds as a product of export and import. In a normal year, the cashew imports to the United States exceeds 50,000,000 lb.

Chestnuts

Currently in North America most of the cultivated chestnuts are of the Chinese variety. Imported chestnuts may be of the European or

Japanese species. For many generations the American chestnut virtually controlled the United States market, but just subsequent to World War I the American chestnut trees which were primarily distributed within the Appalachian mountains were killed by a chestnut blight called *Endothia parasitica*. This completely wiped out the American chestnut. The wood for further commercial consideration of the American chestnut was salvaged for lumber and for the production of tannic acid, even the second generation sprouting from the older tree stumps became reinfected with this fungus and died within a few years.

The Chinese chestnut has taken the place of the American chestnut in US production. The US Dept. of Agr. carried on an extensive campaign introducing Chinese chestnut seed from Georgia to the West Coast and as far north as Maryland and Virginia. The Chinese chestnut has a thin skin and the kernel or meat separates readily. The trees do well in any area but must be protected from the late spring frosts to assure fruiting. The tree itself will withstand low temperatures to −20°F. The young trees will produce heavily even by the tenth year of growth. Fortunately the Chinese chestnut is resistant to the fungus that destroyed the American types.

The nuts vary widely in quality and great care must be taken in curing to maintain the palatability. The nut must not be permitted to dry out completely but must be reduced to 15% moisture to change part of its starch content to sugar. The best cured Chinese chestnuts developed to 5% sugar in the kernel although frequently late cured Chinese chestnuts may show as high as 10% sugar. While the history of Chinese chestnuts production in the United States is not over a sufficiently long enough period to provide accurate data, healthy trees are expected to produce 70 lbs of nuts in the shell by the 15th year of cultivation.

France, Italy, Portugal, Spain and Japan have been importing Chinese chestnuts for the past 40 yr during the period of low gross production of the nuts within the United States. In a normal year the total imported crop will exceed 7,500 tons, mostly from European sources.

Walnuts English (Persian)

These walnuts by heritage might well be named Judean. History records extensive fields in Solomon's gardens at Ethan six miles from Jerusalem. The original stock moved first to Persia, thence to England, to Carpathian Poland and to California. In most western markets they are sold today as "English" but in Asia they are identified as Persian, and in the United States as either "English" or Californian."

Over the intervening centuries, a strain of the Persian walnuts has been acclimated to the protected valleys of the Carpathian mountains in Poland and Austria and these are now finding an acceptable climate to produce in New England and Ontario even at temperatures as low as −30°F. These oriental walnuts produce best on deep and loose soil, preferably many feet above heavy or rocky subsoil. Other than the Carpathian varieties oriental walnuts grow best in warm but not too hot climates, preferably between 80° and 90°F. Excessive cold may kill the English and Californian varieties.

Producing Countries.—The principal producing countries of Persian walnuts are Italy, France, India and the United States. In an average year Italy will produce 25,000 to 30,000 tons of unshelled walnuts; France a similar amount, India 12,000 to 15,000 tons and United States 75,0000 to 80,000 tons. Other countries that are significant producers are Iran, Syria, Turkey, Rumania and Yugoslavia. The principal exporters are the four countries producing the greatest tonnages of nuts in the shell. Of the shelled nuts the heaviest importations to the United States originate in France, Italy, Rumania and Turkey. Other nations that are heavy importers of Persian walnuts are West Germany, Great Britain, Switzerland, Canada and Belgium.

Storage of Persian Walnuts.—Nearly 90% of English or Persian walnuts are marketed as meats. Preparing them for the sheller is a prestorage process. Most walnuts of these varieties contain as high as 45% moisture. Before shelling this moisture content must be reduced greatly to stabilize the kernels against molding, discoloration and rancidity. The optimum kernel moisture is just about 4%. Before shelling, commercial processes may include rehydration over a short period to stabilize the meats.

English walnuts are much more susceptible to rancidity and darkening of kernels than American black walnuts. High humidity, excessive light and heat, and oxygen, destroy their flavor and salability. The most stable condition of the kernels can be maintained with 3.1% moisture. The free world production of Persian walnuts in the shell is approximately 160,000 tons, of this total tonnage about ½ is outside the United States and a similar tonnage within continental United States. The countries outside the United States export about 45% of their production.

Black or American Walnuts

The black walnut is an indigenous nut tree to the deep well-drained well-limed soils of North America; they were found on this continent by the early colonists and had been an annual crop harvested by the American Indians as edible food nuts for many generations. On good soil they are productive within 5 yr after planting. On Eastern and Midwestern United States soils and over the clay land Southern States from Texas to Virginia we find the heavy producing areas of American black walnuts. On account of the rapid growing characteristic of the walnut trees and the high value of walnut wood for furniture manufacture, walnut forestry has been very successful. The annual nut crop contributes very profitably to the maintenance cost of the plantings until the wood can be claimed for valuable sawlogs. As a private venture, the investment return on a walnut forest is not attractive since it usually requires 40 to 50 yr for the complete planting cycle to render full return. Such an investment is more likely to attract a corporation than a private individual.

While there are now available some one hundred varieties of American black walnuts, the nut growers associations are partial to the Stabler, the Thomas and the Ohio varieties. This decision is based upon the productivity, the recovery of nut meats per bushel and the excellency of the product. Some of the one hundred varieties have exceedingly hard shells and small nut meats. In regions where there is some choice even the squirrels prefer to bury the walnut varieties that have hard, thick shells.

Ground walnut shells are used as a substitute for air blasting metals. In engine cylinders the hard wood dust will cut away the carbon but not the metal and serves in burnishing the surfaces cleaned.

Black Walnuts in the Food Industry.—The meats of the black walnut are considered of high value in flavoring ice cream, candies and cakes. In the market place the black walnut growers experience a seller's market since the demand almost annually exceeds the supply. Since the black walnut flavor is somewhat more distinctive than even the pecan, the latter can be considered the nut meat substitute for black walnuts.

Storage of Black Walnuts.—In cold storage held at 70°F or below and 75% RH black walnuts keep well for more than 2 yr without any distinctive flavor loss. The more critical period of their storage is in preparation. It is desirable to remove the hull as soon as it will loosen from the nut. The high percentage of acid-like liquid within the hull will darken the meat if left on the nut too long.

Suitable hullers are available, both hand and power operated. For many years corn shellers of the Midwest were employed to do extra duty in shelling walnuts. Black walnuts should be removed from the hull within two weeks after collecting.

For the market the kernels are classified for (a) large fancy; (b) large medium; (c) regular medium, and (d) small.

Filberts or Hazel Nuts

Hazel nuts are native to North America, Europe and Asia, but do not appear south of the equator. In Midwestern and Eastern United States, they are found indigenous to these areas especially on the edges of uncultivated wooded lands and stock pastures. They usually occur as a tall bush or bushy tree-like shrub.

The European hazel nut under the name of "filbert" has been extensively cultivated as a food supplement and both trees and nuts vary widely in size and productivity. In Asia, especially on the southern Black Sea Coast in Turkey, hazel nuts are indigenous as trees and constitute a profitable export especially in the region of Trapizan in eastern Turkey on the Black Sea. Filbert or hazel nut culture has been practiced for 3,000 to 5,000 yr by the Chinese, the Greeks and the Persians, primarily within the north temperature zones.

North American countries import some 3,500 tons of shelled hazel nuts and 1,200 tons of nuts in the shell annually. The principal imports are from Italy, Turkey and Spain. The United States in turn exports approximately 500 tons of hazel nuts, their principal market being Germany, Canada and South America. The total annual world production is estimated at 150,000 short tons with Italy leading with 70,000 short tons.

In harvesting it is advisable to pick the nuts from the shrub or trees to avoid their falling onto moist or wet ground cover and getting wet. The tannin of the shell penetrates into the nut and imparts a bitter taste. Commercially the freshly picked nut is dried either by the sun or in 100°F driers. These driers appear in many designs; most often the design is a copy of simple evaporated fruit sun driers.

Filberts are stored in the shell in rooms at 65 to 70% RH and a temperature of 32°–35°F. When shelled they keep exceptionally well for 2 yr when stored in carbon dioxide gas even at temperatures up to 50°–70°F.

There are about 1 doz species of filberts indigenous to the temperate zone but they vary widely in type. Northern US "hazel nuts" are bushes about the height of a man but in the eastern area they approach the classification of trees from 25 to 30 ft in height. In Turkey on the Black Sea the filbert tree grows to a height of 60 ft while the Chinese tree varieties even exceed 100 ft.

Peanuts

Peanuts have a much greater market appeal in North America than in Great Britain and Europe. For most Europeans, they are known as ground-nuts. Several attempts have been made by the British to extend their culti-vation to African countries but the great mass of the peanut culture of the world is in North America, especially in the states of: Virginia, North and South Carolina, Georgia, Alabama, Texas and Louisiana. In the United States, the annual crop may exceed 160 million pounds of nuts in the shell. The shell may vary from 30 to 60% of the gross weight in-shell, this variation being due to variety selection.

As harvested, the average crop produces for the world market, shelled nuts that are 44 to 48% oil; 25 to 30% protein; 5 to 7% water and 2 to 4% carbohydrate with smaller quantities of phosphorus, calcium and niacin. Of the edible peanuts harvested in the United States over 50% goes into peanut butter; 22% into peanut candy; 25% into salted peanuts.

For storage they are usually held at 32° to 36°F and 65% RH.

In world production one variety is primarily produced for oil, the other for edible stock. It is estimated that of world production of peanuts 10,000,000 tons go into oil production. In North America, how-ever, more than 65% of the peanuts grown go into the shell trade, either as plain or salted roasted peanuts, peanut butter and peanut candies.

The process of preparing for edible trade includes cleaning, shelling, blanching and roasting, then refrigerated in cold storage at a maximum of 36°F and 65% RH.

Peanuts in Storage.—Bissell and Du Pree (1946) reported that shelled peanuts in heavy cotton bags were free of insect infestation but in jute bags were heavily infested after storage at 80°F and 65% RH. The test room itself was heavily infested with insects in which this observation was made. Thompson *et al.* (1951) found that shelled peanuts in 120 lb burlap bags could be stored for 5 years at 0°F, for 2 years at 32°F, for 1 year at 47°F, and for 3 months at 75°F. All of these tests were at 65% RH.

Mather *et al.* (1956) found that the optimum condition for cold storage of peanuts in gunny sacks was 32°–35°F if held at 85–90% RH. After 9 months of storage under these conditions the shipment was 100% marketable and appeared in excellent condition.

Pecans

Although first grade pecans bring one of the highest prices of any other food nuts on the American market, there is great carelessness in many communities in the collection of the annual crop from the producing trees. This carelessness may be noted from Georgia to Central Texas and into Northern Mexico.

Especially in the regions of high rainfall and heavy grass ground-cover high quality nuts are too often allowed to lie on the damp ground for several days; the result is dark colored nuts with a musty taste.

There are strains of pecans acclimated to upper Ohio and Mississippi Valley but those varieties have small meats and are not of great commercial importance in competition with the Gulf Coast and Southwest varieties.

The US Dept. of Agr. has announced specific instructions on grading and harvesting quality pecans and recently has developed mechanical tree shakers to reduce the time of picking. Woodroof (1967) of Georgia Experiment Station, Experiment, Georgia has devoted much of his two volumes on tree nuts to the grading process and storage of pecans. Any storage house manager who contemplates storing pecans should consult these volumes.

Since the annual volume of exported US pecans exceeds 1.5 million pounds of the shelled crop and a similar weight of in-shell nuts, there is a demand for pecan holding rooms of both the in-transit export and domestic types of both cold and freezer storage. The heaviest importers are Canada, Great Britain, West Germany, Finland, Australia, Japan and South Africa, Sweden, South Arabia, The Netherlands and Switzerland, and Venezuela. The wholesale average price of pecans in-the-shell is over $500 per ton. Of the five principal in-shell nuts produced in large volume in North America only almonds bring a higher average annual price.

Pecans and Pecan Warehousing.—The pecan has been named the most important nut bearing tree in North America. It is a species of the hickory. It is native to southwest United States and northern Mexico. These were planted in this region under Indian culture. The Spaniards found trees of several hundred years production in the sixteenth century, especially along the higher river lands of South Texas and Northern Mexico. Before World War I the American pecan was promoted primarily for local cultivation and regional marketing.

While many orchards existed from Georgia to New Mexico by 1920, widespread cooperative marketing and processing did not become effective in the producing arena until 1930.

Rapid advances have been made in pecan cultivating, harvesting, curing and warehousing during the past half century, while before World War I the pecan was given over to local or area consumption, and was sold

TABLE 21.2

GRADES OF PECANS AS DELIVERED INTO COLD AND FREEZER WAREHOUSES
PECANS-IN-THE-SHELL OFFICIAL STANDARDS CLASSIFICATION
FOR SACKING AND OTHER PACKAGING

Classification	Average Pecan Count per Lb
Oversize	52 or less
Extra large	53 to 60
Large	61–73
Medium	74–90
Small	91–115

US Standard Shelled Pecan Grades:
  (A) US No. 1 halves
  (B) Commercial halves
  (C) US No. 1 pieces
  (D) US Commercial

at a low price per bushel or per pound. The principal assurance of pecan quality to the purchaser of the product was based on the historical record of the established grower to deliver carefully harvested and properly cured nuts in the shell as in previous years.

In this early period most of the shelling and packaging was done in Northern consuming areas. Itinerant or local buyers purchased thousands of bushels in the shell in the Southern growing centers then shipped them to the shellers of the Northeastern United States. These buyers depended upon their own organoleptic inspection and established records of the grower to supply an acceptable product.

Only a limited number of acceptable shelling machines were on the market until 1935. Most corporations engaged in packaging shelled pecans considered their shelling equipment designs as their secret of the trade. By 1940, however, several highly improved shelling machines had been invented for producing acceptable pecan halves at buying centers from San Antonio, Texas to Albany, Georgia and Northward to Memphis, Tennessee and the Tennessee River Valley and Southern Oklahoma and Arkansas.

TABLE 21.3

US STANDARDS SHELLED PECAN HALVES CLASSIFICATION

Grade	Halves per Lb
Mammoth	200–250
Junior mammoth	251–300
Jumbo	301–350
Extra large	351–450
Large	451–550
Medium	551–650
Topper	651–750

TABLE 21.4

AVERAGE GROUND AND TREE NUT PRODUCTION AND PRODUCING AREAS

Common Name	Kind of Plant	Origin	Estimated Total Annual Commercial Production of Principal Producing Countries of Record Thousand Lb	Principal Uses and Producing Areas
Almond	Deciduous tree	Eastern Asia	160,000	Food: Mediterranean countries, United States, China, Iran, Spain, France
Brazil nut	Evergreen tree	South America	70,000	Food: Brazil, Bolivia
Cashew	Evergreen tree	Tropical America	100,000	Food: Commercial oil, India, East Africa
Chestnut	Deciduous tree	Asia Minor, China, Japan	30,000	Food: Mediterranean countries, China, Japan, Europe
English or Persian walnut	Deciduous tree	Asia Minor	70,000 (unshelled)	Food: oil; Mediterranean countries, United States, China, India
Walnuts American black	Deciduous tree	North America	50,000	Food: North America
Filbert	Deciduous shrub or small tree	Asia Minor	300,000 (unshelled)	Food: Turkey, Italy, Spain, United States, Afghanistan
Peanut	Annual plant	South America	200,000 (unshelled)	Edible oil, food: India, Africa, China, United States
Pecan	Deciduous tree	North America Western Europe and Australia	2,500 (unshelled) 1,500	North America, and Europe

Pecans are more sensitive to taste and quality change than walnuts, peanuts or almonds. During the harvesting period, especially that from the tree to the sheller, the quality of the nuts that reach the cold and freezer warehouse will depend largely upon the growers' and brokers' handling. They must pass through the normal curing process, thence to the packing room, and/or the sheller for delivery, to the warehouse for classification for quality for the national market. Even a few hours delay beneath the pecan tree on damp ground may be sufficient to produce an inferior product and a definite price downgrading.

TABLE 21.5

SOME SELECTED STORAGE CONDITIONS OF COMMERCIAL NUTS AND RESULTS[1]

Product and Package	Storage Temperature °F	Storage % RH	Safe Storage Period Months
Almonds, unshelled in cloth bags	32		12
Almonds, shelled—in cans	32	65–75	20
Brazil nuts—shelled	32		6
Cashews	45		12
Chestnuts	32		12
English or Persian walnut, shelled— vacuum packed	32—20		23
American black walnut, shelled—cloth bags	32—20		20
American black walnut, shelled—vacuum packed	32—20		36
Filberts, shelled—polyethylene bags— sealed	32		12
Peanuts, shelled—120 lb burlap bags	0		60
Peanuts, unshelled—gunny sacks	32	85–90	9
Peanuts, shelled and exposed to air	32		6–10
Pecans, shelled—vacuum packed	32–20		24–36
Pecans, shelled—air tight containers	0–20		24–36
Pecans, unshelled—less than 7% moisture in metal cans	0–20		60
Pecans, unshelled—less than 7% moisture, exposed	32–20	75	12

[1] From Woodroof. Foods for shelter storage, Georgia Experiment Station (1960).

Pecan pieces are graded by the dimension of round sieve openings. These sizes are from $\frac{1}{2}$ in. sieve round opening classified as Extra Large to the descending scale openings of Large, $\frac{5}{16}$ in.; Medium, $\frac{1}{4}$ in.; Small, $\frac{3}{16}$ in.; Midget, $\frac{1}{8}$ in.; and Meal, $\frac{1}{16}$ in.

Pecans are of a more distinctive flavor than most other nuts but they are also subject to a higher rate of deterioration than any other of the edible nuts, such as black or Persian walnuts, almonds and peanuts. All nuts keep better, however, when frozen, especially if they have been cured to less than a 6% moisture content.

On account of their high oil and low water content the freezing point of common edible nuts is low. Woodroof shows Persian walnuts begin to freeze at 20°F and pecans at 19.6°F. With low moisture content pecans have a wide freezing range that might extend over 20°F. In other words, while any shipment of pecans may begin to freeze at 19.6°F, it will probably be 0°F before the final moisture portion of each nut is completely solidified or frozen. For above freezing pecan storage, nuts held at 33°F for up to 6 months periods are satisfactory at 65 to 75% RH in controlled atmosphere rooms.

BIBLIOGRAPHY

ANON. 1962. Producing walnuts in Oregon. Cooperative Exten. Serv., Oregon State Univ., Corvallis, Oreg.

ANON. 1965. Serial Rept. Brazil Nut Assoc., New York.

ANON. 1966. Foreign Agr. Serv. Rept. July 18. US Dept. Agr.

BAILEY, L. H. 1949. Manual of Cultivated Plants, Rev. Edition. Macmillan Co., New York.

BISSELL, T. L., and DuPREE, M. 1946. Insects in shelled peanuts. J. Econ. Entomol. 39, 550–552.

BEARD, E. 1962. Hardy nuts. Horticulture 40, 514–515.

BRISON, F. R. 1945. The storage of shelled pecans. Texas Agr. Expt. Sta. Bull. 667, College Station, Tex.

CALIFORNIA ALMOND GROWERS EXCHANGE. 1965. California almonds, varieties, sizes and products. Sacramento, Calif.

DUNCAN, D. 1965. Your filbert control board. Proc. Nut Growers Soc., Corvallis, Oreg.

FAIRCHILD, D. 1930. Exploring for Plants. Macmillan Co., New York.

FISHER, H. H. 1964. A survey of pears, nuts and other fruit clones in the United States. US Dept. Agr. Res. Serv. ARS 34–37–3.

KAINE, M. G., and McQUESTEN, L. M. 1952. Propagation of Plants. Orange Judd, New York.

MATHER, P. B., PRASED, M., and KIRPEL, S. K. 1956. Studies in cold storage of peanuts. J. Sci. Food Agr. 7, 355–360.

McKAY, J. W. 1963. Crane and Orrin, two new Chinese chestnut varieties released by the US Dept. Agr. Fruit Varieties and Horticultural Dig. 17, No. 4, 73–74.

MOLDENKE, H. N., and MOLDENKE, A. L. 1952. Plants of the Bible. Chronica Botanica, Waltham, Mass.

PAINTER, J. H., and RAWLINGS, C. O. 1961. Producing walnuts in Oregon. Oregon Exten. Bull. 795. Oregon S. Exten. Div., Corvallis, Oreg.

POWELL, J. V. 1964. Domestic tree nut industries. US Dept. Agr., Agr. Econ. Rept. 62. Washington, D.C.

REHDER, A. 1956. A Manual of Cultivated Trees and Shrubs in North America, 2nd Edition. Macmillan Co., New York.

RICHARDS, S. I. 1964. Trends in India Agricultural Trade. Agr. Econ. Res. Serv. Rept. 15, US Dept. Agr., Washington, D.C.

SCHREIBER, W. R. 1950. The Amazon Basin Brazil Nut industry. Foreign Agr. Rept. 49. US Dept. Agr., Washington, D.C.

STUCKY, H. P., and KYLE, E. J. 1925. Pecan Production. Macmillan Co., New York.

THOMPSON, H., CECIL, S. R., and WOODROOF, J. C. 1951. Storage of Edible Peanuts. Georgia Expt. Sta. Bull. 268. Experiment, Ga.

WOODROOF, J. G. 1966A. Peanuts: Production, Processing, Products. Avi Publishing Co., Westport, Conn.

WOODROOF, J. G. 1966B. Storage of candies, dried fruits and dried vegetables. In ASHRAE Data Book, Refrigeration Applications. ASHRAE, New York.

WOODROOF, J. G. 1967. Tree Nuts, 2 Vols. Avi Publishing Co., Westport, Conn.

WOODROOF, J. G., and HEATON, E. K. 1953. Year Round on Pecans by refrigerated storage. Food Eng. 25, No. 5, 83–85.

CHAPTER 22

Refrigeration of Furs and Woolens

INTRODUCTION

The large investment in furs within North American and Western Europe, and their susceptibility to moth and beetle damage at high temperatures, has made cold storage a very popular and economic investment for both user and the cold storage operator. What is being said about furs is applicable to woolens except that the high insurance rate levied against furs may not apply to woolens. All such rates are based on commercial values.

Although fumigation methods can be used to control the moths and beetles, it requires more skill to preserve furs effectively by any of the fumigation processes than it does by controlled refrigeration. Both furs and woolens may be given protection in storage by one of the several methods of fumigation. While the fumigation methods of control in killing the moth and larvae is very effective, it does not immunize against reinfestation. Fumigation must be repeated at more-or-less regular intervals for continuous control of the larvae.

Fumigants that are used with much success are: (1) Napthalene in the form of flakes or as commercial moth balls; these are used extensively in closets, trunks, and wardrobes. (2) Parachlorobenzene is used by some small community storage houses especially for woolens. It is as effective as napthalene but, like it, must be repeated at intervals to make it effective over a period of time. (3) Hydrocyanic acid gas is more effective than either napthalene or parachlorobenzene but much more dangerous to humans and to animal life when not expertly applied. (4) Burning sulfur fumes can be most effective as a fumigant for both moths, beetles, and larvae, but such fumes bleach both woolens and furs if strong enough to kill the larvae. (5) Carbon disulfide is an excellent fumigant, but it is highly inflammable and not recommended for use by a novice. (6) Carbon tetrachloride is as effective as carbon disulfide; but instead of being inflammable, it is recognized as a fire extinguisher as well as a most effective fumigant for moths and larvae. (7) A mixture of ethylene dichloride and carbon tetrachloride is very effective in controlling moths, beetles, and larvae, and it is both noninflammable and nonexplosive. To assure both life and fire insurance protection, any enterprise involving fumigants in buildings or vaults should be discussed with the insurance company covering the project and their recommendations considered before proceeding. If inflammable fumigants are to be considered, the operator should keep in mind that any vigorous rubbing of either fur or woolen fabrics in dry rooms may produce a substantial spark of static electricity.

Furs in cold storage may be kept indefinitely without any accelerated deterioration other than the ordinary aging. Relative humidity between 55 and 65% is considered an ideal atmosphere for fur if the storage temperature is held between 35° and 40°F. Before any furs are stored for any length of time, however, the moths, beetles, and larvae should be killed. If furs are stored for 24 hr at 0°F the eggs of the moth will

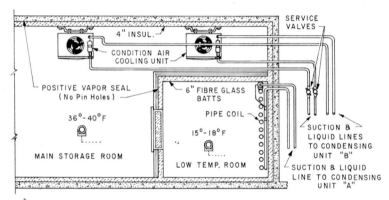

FIG. 22.1. FUR COLD STORAGE WITH MOTH AND BEETLE FREEZING CABINET INSTALLED WITHIN

be completely killed. It will require approximately 4 days for the same results at 15°F. Experimental studies and practices have verified that the best results can be obtained by putting the furs through repeated cycles from the low temperature of 15°F to the warmer temperature of 50°F on a 3- or 4-day holding temperature cycle for 3 or 4 such cycles. This might require a total of 2 to 3 weeks going through this temperature cycle program to completely kill all moths, beetles, and larvae before being moved into final 35°F cold storage.

A program of introducing the furs or woolens into more-or-less permanent storage is to first give them a thorough cleaning and brushing, then freeze them for 4 or 5 days at 15°F, then transfer them to a 50°F cold room at 55 to 65% RH for 2 or 3 days. Then, repeat this cycle for a minimum of 3 times before finally moving the load into a cold storage room held uniformly at 35°F and 55 to 65% RH storage for the term of storage required by the customer.

Experienced managers report that they have kept furs of high value under the above cycle treatment for 20 yr without appreciable deterioration.

PLANT DESIGN

Since, in killing moths and larvae, it is necessary to vary the temperature of the furs or woolens from 50°F at 55 to 65% RH down to 15°F for a period of several days for larvae freezing, the storage space should be varied in temperature over this range and the rooms' vapor sealing, humidity control, and insulation can be effectively varied and maintained as needed over these ranges of air quality and temperature. If, however, it is anticipated that there will be new deliveries of furs for storage at frequent intervals, then it will be more economical of power and refrigeration to build a separate freezer room to periodically carry each load down to 15° to 18°F for the required cycle changes, then there could be a second warmer holding room with a range from 50°F and 55 to 65% RH for the warming period of each load cycle. Finally, this same large room would be held at cold storage temperatures of 35° to 40°F and 55 to 60% RH for the long term holding of the furs over the seasons.

The preferred refrigerant to be used in the compressors for fur storage rooms both for the sharp freezer and the cold storage room is Refrigerant 12. Ammonia is avoided by most all builders of cold and freezer fur plants on account of the danger of injury to the furs of any leaking refrigerant.

Some dimensions that are more-or-less standard for the fur storage rooms are as follows: height of rooms to hold miscellaneous furs in storage

for 1 rack high 7 ft 6 in.
for 2 racks high 10 ft 2 in.
for 3 racks high 13 ft

The space between racks and walls and aisles should be 2.5 ft. The height of each individual rack should be 5.5 ft. A typical rack arrangement is shown in Fig. 22.2. Schematic drawings have been provided herein to show the relation of the freezing room and the storage space. The freezer room is shown in Fig. 22.1 within the cold room to reduce the losses of the cold by infiltration. It will also assist in holding the frost penetration of the freezer room to a minimum.

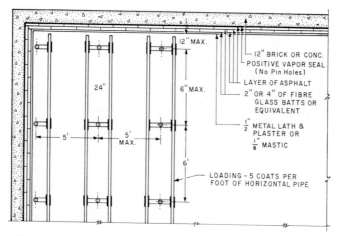

FIG. 22.2. SECTION OF FUR COLD STORAGE ROOM WITH RACK
SUPPORTS LOCATION

Machine Equipment

The principal equipment for the fur storage complex will be the necessary condensing units made up of the compressor, the condenser, and the electric motor. To this will be added the air cooling coils and the evaporator. Each room should be equipped with the necessary circulating fans to move the cooling air in between the several racks loaded with furs.

The defrosting of the freezing coils must be planned in order to eliminate all forms of frost accumulation within the rooms. Moisture infiltration and frost may be caused in several ways. While poor insulation may be partly responsible, the infiltration through cracks and doors is more difficult to curb. All freezer rooms should have a vestibule large enough to accommodate a small fur truck. This will reduce the infiltration on delivery, but even this will not eliminate the customer-introduced warm air through the vestibule.

Sharp freezing rooms at any freezing temperature should never open directly into hot air heated spaces or to the outside. Both the holding and the freezer room should be most adequately vapor sealed to prevent moisture being forced into the rooms by the vapor pressures caused by the wide difference of temperature between the out of doors and the rooms within.

In addition to vapor sealing and insulating of the room for fur storage, the inner finishing coats of plaster or sealing mortars should be pre-dried for several days or weeks to assure no rise in humidity and of frosting by the moisture remaining in the walls and sealants.

Other fumigants that are used with some success are: (1) a mixture of methyl bromide and carbon dioxide; (2) a mixture of methyl formate and carbon dioxide; and (3) a fumigant mixture by volume of 3 parts ethylene dichloride, 1 part carbon tetrachloride, 1 part ethylene oxide, and 9 parts carbon dioxide. The US Dept. of Agr. Circular *369* recommends this mixture be applied in dosages of 15 lb per 1,000 cu ft of space.

TABLE 22.1

TEMPERATURE AND TIME REQUIREMENTS FOR KILLING MOTHS IN STORED CLOTHING

Storage Temp °F	All Eggs Dead After Days	All Larvae Dead After Days	All Adults Dead After Days
0–5	1	2	1
5–10	2	21[1]	1
10–15	4	—	1
15–20	—	—	1
20–25	21	67	4
25–30	21	125[2]	7
30–35	—	283[3]	—

Source: Table AMS-57, US Dept. Agr.
[1] 50–95% of larvae may be killed in two days.
[2] A few larvae survived this period.
[3] Larvae survived this period.

The lesser investment necessary to provide protection to furs by fumigation in preference to the refrigerated cycling down to 15°F and up to 50°F has increased fumigant adoption by the cold storage companies. Where the investment in a freezer vault and equipment for the cycling procedure has been considered too high by the managment, the erection of a fumigation vault has been substituted. When most expertly done, fumigation can be a satisfactory substitute.

RAW PELTRIES

While many furs have been family possessed for 2 or 3 decades, and the number of furs in world-wide use today soars up to the millions, the price of new peltries also mounts upward each year at a rate indicative of their user acceptance.

Peltries from wild animals are obtained from hunters, trappers, and farmers in practically all countries throughout the globe; some domestic animals are pelted when they are a few days old for it is then the hair is short. For the most part, peltries are dried in the open air away from direct sunlight and are packed and shipped to collecting centers for world distribution.

TABLE 22.2

FUR PRODUCTS NAME GUIDE

Name	Family	Name	Family
Alpaca	Camelidae	Leopard	Felidae
Antelope	Bovidae	Llama	Camelidae
Badger	Mustelidae	Lynx	Felidae
Bassarisk	Procyonidae	Marmot	Sciuridae
Bear	Ursidae	Marten, American	Mustelidae
Bear, polar	Ursidae	Marten, baum	Mustelidae
Beaver	Castoridae	Marten, Japanese	Mustelidae
Burunduk	Sciuridae	Marten, stone	Mustelidae
Calf	Bovidae	Mole	Talpidae
Cat, caracal	Felidae	Monkey	Colobidae
Cat, domestic	Felidae	Muskrat	Muridae
Cat, lynx	Felidae	Nutria	Capromyidae
Cat, manul	Felidae	Ocelot	Felidae
Cat, margay	Felidae	Opossum	Didelphiidae
Cat, spotted	Felidae	Opossum, Australian	Phalangeridae
Cat, wild	Felidae	Opossum, ring-tail	Phalangeridae
Cheetah	Felidae	Opossum, South	
Chinchilla	Chinchillidae	American	Didelphiidae
Chipmunk	Sciuridae	Opossum, Water	Didelphiidae
Civet	Viverridae	Otter	Mustelidae
Desman	Talpidae	Otter, sea	Mustelidae
Dog	Canidae	Pahmi	Mustelidae
Ermine	Mustelidae	Panda	Procyonidae
Fisher	Mustelidae	Peschanik	Sciuridae
Fitch	Mustelidae	Pony	Equidae
Fox	Canidae	Rabbit	Leporidae
Fox, black	Canidae	Raccoon	Procyonidae
Fox, blue	Canidae	Raccoon, Asiatic	Canidae
Fox, cross	Canidae	Raccoon, Mexican	Procyonidae
Fox, grey	Canidae	Reindeer	Cervidae
Fox, kit	Canidae	Sable	Mustelidae
Fox, platinum	Canidae	Sable, American	Mustelidae
Fox, red	Canidae	Seal, fur	Otariidae
Fox, silver	Canidae	Seal, hair	Phocidae
Fox, white	Canidae	Seal, rock	Otariidae
Genet	Viverridae	Sheep	Bovidae
Goat	Bovidae	Skunk	Mustelidae
Guanaco, or its		Skunk, spotted	Mustelidae
young, the		Squirrel	Sciuridae
Guanaquito	Camelidae	Squirrel, flying	Sciuridae
Hamster	Cricetidae	Suslik	Sciuridae
Hare	Leporidae	Vicuna	Camdelidae
Jackal	Canidae	Viscacha	Chinchillidae
Jackal, cape	Canidae	Wallaby	Macropodidae
Jaguar	Felidae	Weasel	Mustelidae
Jaguarondi	Felidae	Weasel, Chinese	Mustelidae
Kangaroo	Macropodidae	Weasel, Manchurian	Mustelidae
Kangaroo-rat	Macropodidae	Weasel, Japanese	Mustelidae
Kid	Bovidae	Wolf	Canidae
Kinkajou	Procyonidae	Wolverine	Mustelidae
Koala	Phascolarctidae	Wombat	Vombatidae
Kolinsky	Mustelidae	Woodchuck	Sciuridae
Lamb	Bovidae		

The quality of wild animal peltries is governed by seasonal and sectional differences, and the season producing the best qualities is midwinter. Those taken before and after the full priming season have hair and fur that are not at their best stages of development, and are rated as second and third grade peltries, or as early-caught and late-caught. Spotted animals, such as the leopard and ocelot, are valued by the short length of hair, color of background and shape of rosettes (grouping of spots). Domestic animal peltries are sorted according to the waviness of the hair, length of curl, and beauty of curl pattern. Other factors of valuation are individual size, color, and specialized pattern or curl design.

The animal groups used for furs are rodents, felines, canines, mustelas or weasels, marsupials, ungulates, insectivores, and seals.

Not all fur-bearing animals become prime in the winter season; water rodents such as beaver, nutria, and muskrat attain their best quality during the spring months when the water from the mountain streams is coldest. Other animals, such as the marmot, hibernate during the winter months so that only the peltries of those caught in the fall and spring are available. In the tropics, where the temperatures are quite uniform throughout the year, the animals are taken chiefly during the dry season; and because of the warm climate they may have a poor growth of fur, but for some fur uses such as rugs, mats, and coats short hair furs are in great demand.

Wild fur-bearing animals are protected in most countries by trapping laws which are enforced by penalties of imprisonment and fine. The American fur trade pays little attention to low-grade raw skins, thus discouraging illicit trapping. Fur seals are taken under the supervision of the US Department of the Interior on the Pribilof Islands, shipped to St. Louis, and processed for the account of the US government, which sells them at auction semiannually.

ANIMAL FURS IN THE WORLD MARKET

The commercial value of a fur is based on its points of durability and its beauty. Deterent points would include the inherent odor of the animal from which the fur comes, often an odor that cannot be completely eliminated by processing, and by the tendency of the hide to lose its natural oils and get stiff.

The six marketable furs of greatest durability in alphabetical order are chinchilla, ermine, fox, sea otter, seal, and sable. These six, together with some score of the less valuable furs, are given in Table 22.3. The relative durability and weight in ounces per square foot are tabu-

TABLE 22.3

DURABILITY OF FURS BASED ON A SCALE OF 100

Fur	Points of Durability	Weight in Oz per Sq Ft
Beaver (water hairs cut level with fur)	90	4
Beaver (water hairs removed)	85	3 15/16
Caracal kid	10–15	3 1/4
Chinchilla	15	1 1/2
Ermine	25	1 1/4
Fox		
Tinted black	25	3
Tinted blue	20	3
Silver or black	40	3
White	20	3
Hare	5	1 3/4
Lamb		
Persian lamb	65	3 1/4
Grey lamb	30	3 1/4
Broadtail lamb	15	2 1/4
Lynx, natural	25	2 3/4
Lynx, tinted black	20	2 3/4
Marmot, tinted	10	3
Marten baum, natural	65	2 3/4
Marten baum, tinted	45	2 3/4
Marten, stone	40	2 3/4
Mink	70	3 1/4
Moleskin	7	1 3/4
Musquash, natural	37	3 1/4
Musquash, water hairs removed, sheared and seal finished	33	3 1/4
Nutria	27	3 1/4
Opossum	37	3
Otter		
With water hairs	100	4
With hairs removed	95	3 15/16
Sea (for stoles and collars)	100	4 1/4
Rabbit	5	2 1/4
Sable		
Topped, i.e., top hairs colored	55	2 1/4
Tinted, i.e., fur all colored	50	2 1/2
Sable	60	2 1/2
Seal	75	3
Skunk	70	2 3/4
Squirrel	25	1 3/4

lated therein in alphabetical order irrespective of their ranking for these two features.

MANUFACTURE AND GARMENT CONSTRUCTION OF FURS

Since no two peltries are identical, it is necessary to place each one of them in its proper place in the garment before the cutting operations begin. Little damages which show up in the peltries are removed by making triangular cuts known as tongues and shifting the tongue to cover the damaged spot; the puckering of the skin which results comes out during the wetting and stretching operations which are done later on. The peltries are cut in various ways to suit the particular pattern, the most extensive being a process known as "letting-out." This consists of making diagonal cuts or else V-shaped cuts so that the pelt can be made longer and narrower or shorter and wider. The sewing together of the peltries is done on a machine that has two rapidly revolving cog wheels with a shuttle and needle that make an overcast stitch. The coat sections are given to a "nailer" who wets the skin side thoroughly and nails the coat sections, consisting of the body, sleeves, and collar, to a soft wood board to conform to the pattern which is chalked on the board. These sections are allowed to dry out thoroughly and are then assembled, the unfinished garment is softened again by putting it into a revolving drum after which the lining and other findings are attached to the coat. The garment is finally given a glazing, which means bringing out the natural luster of the hair. This is done by brushing water on the hair and carefully beating and combing it; in some instances, a lukewarm iron is used to facilitate the straightening of the hair, much as in the pressing out of textiles.

NAME GUIDE TO THE VALUE OF THE FUR PRODUCTS

In any publicity of marketing all the publicity must always contain the true name of the animal from which the fur has come. The US Federal Trade Commission, Bureau of Textiles and Furs, Washington, D.C., has available the Rules and Regulations under the Fur Products Labeling Act; and any processor, buyer, seller, or sales agent should learn the rules and regulations under which he must work within the United States if he is expected to label, invoice, or advertise furs, both new and used.

The Act specifies that it covers any fur product which "is made in whole or in part of fur which has been shipped and received in commerce." A fur cold storage manager should review the contents of these rules and regulations so that he will be cognizant of the conditions of sale, display, and naming of the furs that may be stored and distributed under his roof.

This comprehensive list of fur names recognized in commercial exchange by the Federal Trade Commission, Bureau of Textiles and Furs, includes all of those in which the US Fur Industry has a commercial interest. It does not include furs that are going to be processed as leather for which the manufacturing and the marketing treatment widely deviates from that for commercial furs. As the animal skin products fan out into a very wide array of items from jewelry mounts, to Western saddles, high-grade luggage; also a wide array of footware, dolls, shoes, and heavy military boots. The arts and crafts procedures and designs that have dominated these creative fields for centuries usually prevails.

IMPORT RULE 2 OF US FEDERAL TRADE COMMISSION

Rules and Regulations under the Fur Products Labeling Act, as amended May 15th., 1961. On General Requirements of Fur Imports:

(a) Each and every fur product, except those under $7.00 value, shall be labeled and invoiced in conformity with the requirements of the Act and Rules and Regulations.

(b) Each and every fur shall be invoiced in conformity with the requirements of the Act and Rules and Regulations.

(c) Any advertising of fur products or furs shall be in conformity with the requirements of the Act and Rules and Regulations. (16 CFR 301.2)

BIBLIOGRAPHY

AMERICAN SOCIETY OF REFRIGERATING ENGINEERS. 1957. Fur storage. Air Conditioning, Refrigerating Data Book, 6th Edition. New York.

ANON. 1930. Fur Storage, General Electric Corp., Schenectady, New York.

ANON. 1934. Fur Storage, Frigidaire Corp., Dayton, Ohio.

ANON. 1940. Fur Storage, Carrier Air Conditioning and Refrigeration Corp., Syracuse, New York.

BACK, E. A., and COTTON, R. T. 1927. Effect of cold storage upon clothes moths. Refrig. Eng. 13, No. 6, 365–366.

MACINTIRE, H. J. 1928. Handbook of Mechanical Refrigeration. John Wiley & Sons, New York.

NATIONAL BOARD OF FIRE UNDERWRITERS. 1947. Standards of the National Board of Fire Underwriters for Fur Storage. Pamphlet 81.

WOOLRICH, W. R. 1966. Handbook of Refrigerating Engineering, 4th Edition. Vol. 2. Avi Publishing Co., Westport, Conn.

WOOLRICH, W. R., and BARTLETT, L. 1948. Handbook of Refrigerating Engineering. D. Van Nostrand Co., Princeton, N.J.

CHAPTER 23

Specialized Storage and Process Rooms

INTRODUCTION

The greatest percentage of products stored in public cold storages can be stored under standard conditions maintained by the warehouse in the various cold rooms. As a rule, the larger warehouses will have several areas of different temperatures which will accommodate most commercially stored products. There are some products, however, that need special treatment and are handled in sufficient volume to justify a specialized room. Some of these specialized conditions will occur often enough to merit mention in some detail.

TEMPERING ROOMS

There are some products stored at cooler temperatures that are normally shipped out of the warehouse in nonrefrigerated transport. Canned milk and a few other canned and packaged products fall into this category. Removing canned or packaged merchandise from cold storage and shipping at high outside summer ambient temperatures can pose a problem whenever the dew point of the outside air is higher than the product temperature. Sweating will occur on the outside surface of the product package which can ruin labels and cause other damage. To prevent this damage, a tempering room is required in which products can be warmed at low humidity conditions to prevent condensation. In any large cold storage operation, this room can be very versatile and useful to the operation of the plant.

A tempering room is usually a regular cold storage room as far as insulation and vapor sealing are concerned. Controls and equipment, however, are somewhat more elaborate than a normal cold storage room would use. Extra surface is required in the cooling units and a source of heat is also required. Direct expansion units with a suction temperature as low as $0°F$ are desirable to give a wide range of control. Suction pressure on the refrigeration units may be varied by means of a modulating back pressure regulator controlled by a humidistat. To raise the temperature of the room and product, a heating coil is required. This coil is best served by a source of hot water in the range of $160°$ to $180°F$ and a motorized valve controlled by a thermostat can be used

286

as a control. Thus, the dry bulb setting is controlled by the heater coil and the humidity by the cooling coil. As lowered humidities are called for, as at the start of a tempering cycle, the cooling coil operates under a lowered (colder) suction to condense moisture from the air. To maintain dry bulb temperature, the heater coil is operated to offset the cooling so that the resultant air in the room is warm and dry. The heating coil must be large enough to overcome the capacity of the cooling coils plus enough capacity to warm the room. Usually a chart can be prepared giving various humidity-temperature-dew point relationships so that the controls can be set for the desired results. Products are normally tempered to final desired conditions over a period of a couple of days. Temperature control of the room is usually raised in several steps over the tempering period. A thermometer placed in the product near the center of the stack should be used as a criterion of when the product may be shipped. The function of the tempering is to raise the product temperature above the highest outside anticipated dew point temperature.

Cooling units should be equipped with some method of defrost since at the start of the tempering process, the refrigerant temperature may be quite low and considerable frost can accumulate on the coil surface. As the temperature is raised, the refrigerant temperature will normally also rise and will eventually be above freezing so that liquid condensate will occur at that time without the necessity of defrosting.

The heating coil may be a separate unit with its own fan. It is not necessary that it be incorporated in the actual cooling unit. The dry cold air from the refrigeration unit will mix very rapidly with the warm air from the heating unit and the dew point conditions in the room will be the same at all points in the room since the air mix will be almost instantaneous. Separate heating and cooling units can be mounted along one wall of the room. If they are ceiling units, they can blow out into the room at ceiling level with no duct work. If they are floor units, a gooseneck duct to carry air up for discharge at ceiling level is usually all that is required.

Since tempering rooms are operated at low humidity conditions, care should be taken that the room insulation is tightly sealed and also that the doors have a tight seal when closed so that excess moisture will not flow into the room.

HIGH HUMIDITY ROOMS

Tempering rooms can also be used for high humidity rooms by reversing controls. Since these rooms are equipped with more than normal cooling surface, a very close temperature difference can be maintained between the room temperature and the refrigerant temperature and very little moisture will be removed by the refrigeration. With a modulating thermostat instead of a humidistat controlling the modulating back pressure valve, any set temperature can be maintained with the highest possible evaporating temperature of the refrigerant. For a variety of humidity settings, the heating coil may be controlled by a humidistat and almost any degree of humidity may be maintained under cooler conditions.

The versatility of a tempering room makes it a very desirable addition to any medium or large cold storage plant. Size can be whatever is desired. A good size range will be with a capacity to hold from 5 to 15 or even 20 carloads of merchandise. Location of the room near the machine room is desirable from the standpoint of being near a source of heat and also convenient for frequent checking by machine room personnel when required. A simple control panel can be located at a convenient point with proper switches to control operation under any category desired. It is also desirable to add temperature and humidity indicators to the control panel.

CANDY AND SPECIAL HUMIDITY ROOMS

Various types of candies are stored in cooler rooms. Temperatures required will vary depending on what the candy manufacturer specifies and also on the type of candy being stored. Many types of candy are stored after manufacture during the year to be drawn out during holiday seasons when demand is high. Humidities are usually low and require rather exacting control. Recording temperature and humidity indicators are very often required to assure continuous supervision. Many types of candy require holding in a room at a temperature near 50°F and about 50% RH. Hard candies are kept at various specified temperatures and humidities to prevent them from becoming sticky or discolored. Chocolate candy also has a tendency to lighten in color and to become color streaked if not held within certain temperature limits. Almost all candy requires a humidity somewhat lower than can be accomplished with cooling coils alone.

To obtain the low humidity required in a candy room, some type of reheat is usually required. In some warehouses candy storage is a sizeable load and it is desirable to have space for this type of storage. The simplest form of refrigeration for a candy room consists of a cooling fan-coil unit with an electric heater mounted on the face of the unit.

Temperature control is obtained by a thermostat operating the cooling coil and humidity is controlled by a humidistat operating the reheat coil. Control and operation is similar to a tempering room except that the requirements are not as severe and control need not be as elaborate. Several cooling coils with reheat mounted along the side of the room at ceiling level or floor units with gooseneck ducts to ceiling level make a good room that can be controlled easily.

In areas of high electric cost, electrical heating for reheat may become quite expensive to operate and other sources of heat become desirable. Quite often, even in large warehouses, the candy rooms may be operated from a separate refrigeration unit since quite high refrigerant temperatures are sometimes used in comparison to the standard cold storage refrigerant temperatures. In these instances, or where the candy room is close to the machine room, a source of heat may be at hand from the compressor discharge gas. If a complete self-contained system is used, condensing coils can be mounted on the face of the cooling units and by means of solenoid valves, hot gas can be diverted from the regular condensers to the coils on the face of the cooling units as a reheat source. By splitting the reheat condensing coils into several circuits or multiple units, step control can be obtained. Of course, hot water or steam may also be used for reheat, if available. Steam is probably the least desirable source since the temperature level is high for reheat.

Film and paper storage rooms also require a rather high degree of humidity and temperature control similar to that of the candy room although the dry bulb temperature may be somewhat lower than for candy. Construction of these rooms is similar to the candy rooms.

CONSTRUCTION AND OPERATION

In all specialized rooms where humidity is held at variance from the outside or natural humidity, careful sealing of the rooms and door openings is mandatory and good vapor barriers are a necessity in the room construction since rather great vapor pressure differences between inside and outside will be encountered. Where a room is consistently used at a low humidity and relatively high temperature (50°F), it is sometimes advisable to use somewhat less than the normal thickness of insulation in the ceiling and walls so that there will be an increased sensible heat load to allow the refrigeration to operate to dehumidify the room. This will tend to lessen the auxiliary heating required. Where outside temperatures reach and hold low levels for extended periods of time, considerable amounts of auxiliary heat may have to be added to obtain a humidity balance. In cold regions, the insulation thickness used in the room should not be decreased.

In low humidity storages, merchandise movement is not rapid and loads are rather stable over long periods of time. In such instances, manual setting of controls can sometimes be used since load conditions will not vary or change except slowly especially in large rooms. Careful and frequent checking is normally required for manual control conditions and savings in control equipment cost are usually more than offset by increased labor cost.

STORAGE OF MISCELLANEOUS PRODUCTS

Edible Nuts[1]

Storage of edible nuts at cooler temperatures has been done for many years. A few years ago the biggest storage load of edible nuts was probably shelled peanuts. They were stored at temperatures of 32° to 35°F and 80 to 85% RH. The shelled nuts were stored in bags weighing 120 lb. Peanuts are easily stored and are not seriously affected in the presence of refrigerant gases. Refrigerant leaks in direct expansion coils will not change the grade of peanuts normally unless actual liquid refrigerant burns the nut.

Pecans are very susceptible to ammonia vapors in the cooler storage room and for this reason, most pecan storage rooms are chilled with brine in plants where ammonia is the prime refrigerant. Pecans are stored in coolers at the same temperature conditions as peanuts but are usually stored in the shell. Pecan storage assumes rather large proportions in the south and southwest.

Recently, both peanuts and pecans have been held in freezer storage. The storage life of these nuts in the shell is about five times as long as can be attained in a cooler. More pecans are held in freezer storage than peanuts due to the fact that the unit price is much higher and greater profits accrue when they are held and sold in a year when the fresh crop is small and prices are high. Pecans are susceptible to ammonia vapor in the frozen state so that brine refrigeration units are usually used in ammonia plants. The presence of ammonia tends to blacken the pecan meat. This does not alter the flavor but the off color characteristics will often change the grade downward making the nuts less valuable.

[1] For further recommendations of edible nut cold and freezer storage see Chap. 21.

Pecans or peanuts may be frozen in an ordinary freezer holding room. Fast freezing of the product is not essential so that only a moderate amount of excess refrigeration over the amount required for holding only is required. After the nuts are frozen, minimum storage tonnage is required since nut rooms are stacked high and the nuts occupy a great deal of the total cubage of the room. In some instances where a long hold is required, pecans may be hand stacked to obtain greater stored tonnage in a given area. This is true in rooms with high ceilings where it is desired to utilize maximum space and the storage period is long and the lots of nuts are large. Local labor costs will determine the feasibility of hand stacking.

Potatoes

The American public has an abundant annual supply of both Irish and sweet potatoes. Both types originated in Central and South America. The Irish name became appended to the now principal staple vegetable food of Western Europe and North America when it was introduced to Ireland to meet the repeated threats of a famine on that island. The threat was met so successfully that since that period, the tuber which more properly should be named Peru potato is called Irish.

The sweet potato is also of an Indian (probably Inca) origin of Central and South America and appears in many variations. The most popular of which is the yam. The yam variety alone of the sweet potato family can boast 150 species.

Storage.—While the annual production of the "Irish" potato in North America alone totals over a half billion bushels annually and the sweet potato not over 50,000,000 bu since both are harvested in the late summer and fall, there is a great demand for storage of several months duration to level out the delivery to the grocers and to the processors of many types of human foods that appear on the American market.

The approximate analysis of White Irish potatoes is given below:

	%
Water	65 to 85
Protein	1 to 4
Carbohydrate	15 to 30
Solids	12 to 35

Potatoes change in many characteristics in storage especially if not well-cured after digging. This will include maintaining the storage room at 50° to 60°F and 85% RH to prevent dehydration. If properly cured their weight loss is under 5% in the first month and 9% in 6 months.

Sweet potato storage usually requires a separate room both for curing and for long-term storage. Most commercial brokers of yam or other sweet potatoes will specify the temperature and humidity storage they anticipate necessary to preserve their shipments.

Many early growers of Irish potatoes tried to avoid the extra cost of commercial warehousing and stored their year's crop on a selected dry ground area, then covered the accumulated pile of potatoes with burlap blankets. The entire pile was then covered with 6 to 8 in. of sandy field soil to protect it from early frost and fall showers. Such storage in the hands of competent, experienced growers was often very successful. But even the best of conditions was subject to unanticipated heavy rainfalls and subsequent freezing weather even to the concurrent fall of snow, with subsequent ice formation.

BIBLIOGRAPHY

GONTHER, R. C. 1957. Refrigerated Air Conditioning and Cold Storage. Chilton Co., Philadelphia, Pa.

SMITH, O. 1968. Potatoes: Production, Storing, Processing. Avi Publishing Co., Westport, Conn.

STANEAR, W. 1950. Plant Engineering Handbook. McGraw-Hill Book Co., New York.

TRESSLER, D. K., VAN ARSDEL, W. B., and COPLEY, M. J. The Freezing Preservation of Foods, 4th Edition, 4 Vols. Avi Publishing Co., Westport, Conn.

WOODROOF, J. G. 1967. Tree Nuts: Production, Processing, Products, 2 Vols. Avi Publishing Co., Westport, Conn.

WOOLRICH, W. R. 1965. Handbook of Refrigerating Engineering, 4th Edition, 2 Vols. Avi Publishing Co., Westport, Conn.

CHAPTER 24

Safety of Workmen in Cold and Freezer Storage Rooms

INTRODUCTION

More than 60% of the accidents occurring in cold and freezer storage including transport and refrigerated processing are preventable. The management of cold and freezer storage warehouses and the related facilities, including the superintendents of the transport and refrigerated processing facilities, can prevent probably 50% of the industrial hazards by judicious planning and design. The other 50% can be charged to the ignorance and carelessness of the employees and the related service group's transporting, handling and processing the cold and freezer storage items of modern storage warehouses.

The management and supervisory personnel are obligated to make all equipment, facilities and service arrangements as foolproof as possible, but the employees must serve the management in good faith and not bring on themselves unnecessary injury to earn a vacation or to receive insurance compensation illegally. Eternal vigil of machine builders, warehouse managers and all employees of the cold and freezer warehouse facilities is essential if accidents are to be maintained at a minimum.

SOME GENERAL AND ADVISORY INSTRUCTIONS

All moving machines should receive careful inspection at scheduled periods, in addition to the casual daily inspection by the operating staff. Gears should be completely enclosed, set screws counter-sunk, power belts completely guarded and open vats and wells fenced or covered. Emery wheel guards should be built of steel plate and so installed as to discourage removal of the guard. Emery wheel operation is not too safe even when all precautions are taken, but a completely steel-guarded wheel and approved grinding goggles on every operator should be required in every plant.

Crushing of the hands or feet by blocks of ice or boxes of produce is

293

a common accident in refrigeration plants. Repeated educational talks to the operators are necessary to reduce this hazard to a minimum.

In ice plants slippery floors cause many falls and injuries. Non-slip shoes are recommended.

Severe electrical shock, often fatal, is common in carelessly managed plants. All lamp cords and all switches and electrical connections should be examined frequently. Especially where moisture and brines are spilled on the floors, fatal grounding through the human body is very possible when exposed electrical connections can be reached by the operator's hands or touched by his head. Great care should be exercised in working with toxic refrigerants, especially in machinery rooms where there is danger of a rupture of a valve or a cylinder head. Some of the worst accidents recorded in refrigeration history have been brought about by careless or uninformed operators not understanding the hazards inherent in starting or stopping a compressor operating on a toxic refrigerant. Compressor rooms should be designed to offer every possibility of the operating force making a quick getaway if a rupture of any part of the compression system occurs.

A nonreturn valve installed in the discharge line is a very good safety precaution against the failure of a cylinder head or machine valve. Remote control stop valves on each compressor are also very desirable for the safety of the compressor room operators. Gas helmets are too often in an inconvenient place. They are a most important part of any plant using toxic refrigerants. An occasional safety drill of the compressor plant personnel is recommended. This will indicate to all concerned the location of the several safety devices and emphasize any weakness in the planned safe practice procedure.

Ice crushers have become notorious for their maiming of operators. If an ice crusher cannot be installed to be operated with safety, then its installation should be deferred until the proper safety appliances can be secured and erected.

Liquid Level Gage Glasses except those of the bull's-eye or reflex type, should have automatic closing shut-off valves, and such glasses should be adequately protected against injury.

A.S.A. CODE ON REFRIGERANT-CONTAINING PRESSURE VESSELS

Refrigerant-Containing Pressure Vessels Exceeding 6 In. Inside Diameter except those having a maximum allowable internal or external working pressure of 15 psig or less, shall comply with the rules of Section VIII of the 1956 Edition of the ASME Boiler and Pressure Vessel Code covering the requirements for the design, fabrication and inspection during construction of unfired pressure vessels.

Refrigerant-Containing Pressure Vessels Not Exceeding Inside Diameter of 6 in. irrespective of pressure, shall be listed either individually or as part of refrigeration equipment, by an approved nationally recognized testing laboratory having a follow-up inspection service. Vessels not so listed shall be constructed according to paragraph 11.1 and 11.1.1 of that code.

A.S.A. INSTRUCTIONS FOR OPERATING ROOM

Substitution of Kind of Refrigerant in a system shall not be made without the permission of the approving authority, the user, and the makers of the original equipment, and due observance of safety requirements, including:

(a) the effects of the substituted refrigerant on materials in the system;

(b) the possibility of overloading the liquid receiver which should not be more than 80% full of liquid;

(c) the liability of exceeding motors horsepower, design working pressure, or any other element that would violate any of the provisions of this code;

(d) the proper size of refrigerant controls;

(e) the effect on the operation and setting of safety devices;

(f) the possible hazards created by mixture of the original and the substituted refrigerant; and

(g) effect of the classification of the refrigerant as provided in this standard.

THOSE WHO HAVE IMMEDIATE NEED FOR EMERGENCY INSTRUCTION

Every user of refrigeration is advised to obtain a copy of the complete code on mechanical refrigeration, B9-1 of the American Standards Association, or ASRE Standard 15–58. For those who need immediate information from the American Standards Association Code B9–1 or ASRE Standard 15–58 on safety; Sections 5, 6, 7, 9, 14, and 15 are presented herewith:

Section 5—Refrigerant Classification

TABLE 24.1

CHEMICAL FORMULAS OF REFRIGERANTS

Group I	
Carbon dioxide (Refrigerant 744)	CO_2
Dichlorodifluoromethane (Refrigerant 12)	CCl_2F_2
Dichlorodifluoromethane, 73.8%	CCl_2F_2
and ethylidene, 26.2%	CH_3CHF_2
(Refrigerant 500)	
Dichloromethane (Methylene Chloride)	CH_2Cl_2
(Refrigerant 30)	
Dichloromonofluoromethane (Refrigerant 21)	$CHCl_2F$
Dichlorotetrafluoroethane (Refrigerant 114)	$C_2Cl_2F_4$
Monochlorodifluoromethane (Refrigerant 22)	$CHClF_2$
Monochlorotrifluoromethane (Refrigerant 13)	$CClF_3$
Trichloromonofluoromethane (Refrigerant 11)	CCl_3F
Trichlorotrifluoroethane (Refrigerant 113)	$C_2Cl_3F_3$
Group II	
Ammonia	NH_3
Dichloroethylene	$C_2H_2Cl_2$
Ethyl chloride	C_2H_5Cl
Methyl chloride	CH_3Cl
Methyl formate	$HCOOCH_3$
Sulfur dioxide	SO_2
Group III	
Butane	C_4H_{10}
Ethane	C_2H_6
Ethylene	C_2H_4
Isobutane	$(CH_3)_3CH$
Propane	C_3H_8

Group I has the greater usefulness because these refrigerants possess low toxicity, explosiveness and flammability.

Group II is next in preference, while Group III refrigerants must be handled with the most discretion and caution.

Section 6—Institutional Occupancies

No refrigerating system shall be installed in or on public stairways, hallways, lobbies, entrances or exits.

Refrigerant piping or tubing shall not be carried through floors except that, for the purpose of connecting to a condenser on the roof, it may be carried through a continuous, rigid and tight fire-resisting (4 hr rating) flue or shaft having no openings on intermediate floors, or it may be carried on the outer wall of the building provided it is not located in an air shaft, closed court or in other similar open spaces enclosed within the outer walls of the building.

Group 1 Refrigerants.—No refrigerating system shall be installed in any room except Unit Systems each containing not more than 10 lb of a Group 1 refrigerant, and then only when a window or other ventilation is provided, and except as otherwise permitted.

Systems each containing not more than 20 lb of a Group 1 refrigerant may be installed in kitchens, laboratories, and mortuaries.

Systems each containing more than 20 lb of a Group 1 refrigerant shall be of the Indirect type with all refrigerant containing parts, excepting parts installed outside the building, installed in a machinery room, used for no other purpose and in which for Group 1 refrigerants, excepting carbon dioxide, no flame is present or apparatus to produce a flame is installed. When a Group 1 refrigerant, other than carbon dioxide, is used in a system where there is an apparatus for producing an open flame, then such refrigerant shall be classed in Group 2 unless the flame producing apparatus is provided with a hood and flue capable of removing the products of combustion to the open air. Flames by matches, cigarette lighters, small alcohol lamps and similar devices shall not be considered as open flames.

Group 2 Refrigerants.—Group 2 refrigerants shall not be used except in Unit Systems containing not more than 6 lb of refrigerant when installed in kitchens, laboratories, or mortuaries, or except in systems containing not more than 500 lb of refrigerant and having all refrigerant containing parts installed in a Class T machinery room. Group 2 refrigerants shall not be used in a system for air conditioning for human comfort, except in an Indirect Vented Closed Surface System, or in a Double Indirect Vented Open Spray System, or in an Indirect Absorptive Brine System.

Group 3 Refrigerants.—Group 3 refrigerants shall not be used in Institutional Occupancies.

Section 7—Public Assembly Occupancies

Refrigerant piping or tubing shall not be carried through floors except from basements to the first floor or from the top floor to a penthouse or the roof, or except that for the purpose of connecting to a condenser on the roof it may be carried through a continuous, rigid and tight fire-resisting (4 hr rating) flue or shaft having no openings or intermediate floors, it may be carried on the outer wall of the building provided it is not located in an air shaft, closed court or in other similar open spaces enclosed within the outer walls of the building.

Group 1 Refrigerants.—The maximum quantity of a Group 1 refrigerant in a Direct System used for air conditioning for human comfort shall be limited by the volume of the space to be air conditioned as shown in the table.

A system containing more than 50 lb of a Group 1 refrigerant, other than carbon dioxide, and which includes air ducts shall be of the Indirect type unless it conforms to requirements as follows:

(a) Positive automatic fire damper or dampers shall be provided to cut off the refrigerant containing apparatus from the duct system.

(b) Automatic means shall be provided to close the dampers and to stop the fan when the temperature of the air in the duct at the damper location reaches 210°F when the damper is on the discharge side of a system containing a heating coil at 125°F when the damper is on the suction side of the system, or on the discharge side of a system containing no heating coil.

A system containing more than 1,000 lbs of a Group 1 refrigerant shall be of the Indirect type with all the refrigerant containing parts, excepting parts mounted outside the building, installed in a machinery room used for no other purpose and in which for Group 1 refrigerants, excepting carbon dioxide, no flame is present or apparatus to produce a flame is installed.

When a Group 1 refrigerant, other than carbon dioxide, is used in a system, any portion of which is in a room where there is an apparatus for producing an open flame, then such refrigerant shall be classed in Group 2 unless the flame producing apparatus is provided with a hood and flue capable of removing the products of combustion to the open air. Flames by matches, cigarette lighters, small alcohol lamps and similar devices, shall not be considered as open flames.

TABLE 24.2

MAXIMUM LEVELS OF GROUP 1 REFRIGERANTS

Group 1 Refrigerant Name	Chemical Formula	Maximum Quantity in Lb per 1,000 Cu Ft of Air Conditioned Space[1]
Carbon dioxide	CO_2	12
Dichlorodifluoromethane (R 12)	CCl_2F_2	30
Dichloromethane (Carrene No. 1)	CH_2Cl_2	6
Dichloromonofluoromethane (R 21)	$CHCl_2F$	13
Dichlorotetrafluoroethane (R 114)	$C_2Cl_2F_4$	40
Trichloromonofluoromethane (R 11)	CCl_3F	35

[1] When the refrigerant containing parts of a system are located in one or more enclosed spaces, the cubical contents of the smallest enclosed space other than the machinery room, shall be used to determine the permissible quantity of refrigerant in the system.

When the evaporator is located in a duct system, the cubical content of the smallest enclosed space served by the duct system shall be used to determine the permissible quantity of refrigerant in the system unless the airflow to any enclosed space served by the duct system cannot be reduced below one-fourth of its maximum, in which case the cubical contents of the entire space served by the duct system shall be used to determine the permissible quantity of refrigerant in the system.

Group 2 Refrigerants.—Group 2 refrigerants shall not be used except in Unit Systems containing not more than 12 lb of refrigerant, or except in systems containing not more than 1,000 lb of refrigerant and having all refrigerant containing parts installed in a Class T machinery room.

Group 3 Refrigerants.—Group 3 Refrigerants shall not be used in Public Assembly occupancies.

Section 9—Commercial Occupancies

No refrigerating system shall be installed in public hallways, lobbies, entrances, or exits, except unit systems containing not more than 4 lb of a refrigerant, provided free passage is not obstructed. Refrigerant piping shall not be carried through floors except as follows:

(a) It may be carried from basements to the first floor or from the top floor to a penthouse, or the roof.

(b) For the purpose of connecting to a condenser on the roof it may be carried through an approved, rigid and tight continuous fire-resisting flue or shaft having no openings on intermediate floors, or it may be carried on the outer wall of the building provided it is not located in an air shaft, closed court, or in some other similar open spaces enclosed within the outer walls of the building.

(c) In systems containing Group 1 refrigerants and used for air conditioning for human comfort, the refrigerant piping may be carried through floors provided it is enclosed in an approved, rigid and tight continuous fire-resisting flue or shaft where it passes through any intermediate space not served by the air conditioning system. The flue shall be vented to the outside or to a space served by the air conditioning system. Such systems shall conform to the requirements of paragraph 9.20.

Group 1 Refrigerants.—Direct Systems containing more than 20 lb of a Group 1 refrigerant, when used for air conditioning for human comfort, shall be limited by the volume of the space to be air conditioned as follows: A system containing more than 50 lb of Group 1 refrigerant, other than carbon dioxide, and which includes air ducts shall be of the Indirect type unless it conforms to the requirements as follows:

(a) Positive automatic fire damper or dampers shall be provided to cut off the refrigerant containing apparatus from the duct system.

(b) Automatic means shall be provided to close the dampers and to stop the fan when the temperature of the air in the duct at the damper location reaches 210 F when the damper is on the discharge side of a system containing a heating coil and at 125 F when the damper is on the suction side of the system or on the discharge side of a system containing no heating coil.

Group 2 Refrigerants.—A system containing more than 20 lb of a Group 2 refrigerant shall not be used for air conditioning for human comfort unless it is of the Indirect Vented Closed Surface, Double Indirect Vented Open Spray, Indirect Absorptive Brine, or primary circuit of a Double Refrigerant type with all the refrigerant containing parts, excepting parts mounted outside the building, installed in a machinery room used for no other purpose.

Any system containing more than 600 lb of a Group 2 refrigerant shall have all refrigerant containing parts installed in a Class T machinery room.

Group 3 Refrigerants.—Group 3 refrigerants shall not be used except in a Unit System containing more than 6 lb of refrigerant.

Section 14—Tests

Every refrigerant containing part of every system that is erected on the premises except compressor, safety devices, pressure gages, and control mechanisms, that are factory tested, shall be tested and proved tight after complete installation and before operation at not less than the minimum pressures shown in Table 24.3 Test Medium. No oxygen or any combustible gas or combustible mixture of gases shall be used for testing.

Refrigerants Not Listed.—For refrigerants not listed in Table 24.3 the test pressure for the high pressure side shall be not less than the saturated vapor pressure of the refrigerant at 150°F. The test pressure for the low pressure side shall be not less than the saturated vapor pressure of the refrigerant at 115°F. In no case shall the test pressure be less than 30 psi by gage.

TABLE 24.3

TEST PRESSURES

Refrigerant		Min. Test Pressure, Psi	
Name	Chemical Formula	High Pressure Side	Low Pressure Side
Ammonia	NH_3	300	150
Butane	C_4H_{10}	90	50
Carbon dioxide	CO_2	1500	1000
Dichlorodifluoromethane (Refrigerant 12)	CCl_2F_2	235	145
Dichlorotetrafluoroethane (Refrigerant 114)	$C_2Cl_2F_1$	80	50
Dichloromethane (Carrene No. 1) (Methylene chloride)	CH_2Cl_2	30	30
Dichloromonofluoromethane (Refrigerant 21)	$CHCl_2F$	70	50
Dichloroethylene	$C_2H_2Cl_2$	30	30
Ethane	C_2H_6	1100	600
Ethyl chloride	C_2H_5Cl	60	50
Isobutane	$(CH_3)_3CH$	130	75
Methyl chloride	CH_3Cl	215	125
Methyl formate	$HCOOCH_3$	50	50
Propane	C_3H_8	325	210
Sulfur dioxide	SO_2	170	95
Trichloromonofluoromethane (Refrigerant 11)	CCl_3F	50	50

For foreign installations, contractors should determine if the US Safety Codes are acceptable for refrigerants and equipment.

Posting of Tests.—A dated declaration of test, signed by the installer, shall be mounted in a frame, protected by glass, and posted in the machinery room. If an inspector is present at the tests, he shall also sign the declaration.

Section 15—Instructions

User's Responsibility.—All refrigerating systems shall be maintained in a clean manner, free from accumulations of oily dirt, waste, and other debris, and shall be kept readily accessible at all times.

Instructions.—It shall be the duty of the person in charge of the premises on which a refrigerating system containing more than 20 lb of refrigerant is installed, to place a card conspiciously as near as practicable to the refrigerant condensing unit giving directions for the operation of the system, including precautions to be observed in case of a breakdown or leak as follows: (a) instructions for shutting down the system in case of emergency; (b) the name, address and day and night telephone numbers for obtaining service; and (c) the name, address and telephone number of the municipal inspection department having jurisdiction and instructions to notify said department immediately in case of emergency.

Signs.—Each Refrigerating System shall be provided with an easily legible metal sign permanently attached and easily accessible, indicating thereon the name and address of the manufacturer or installer, the kind and total number of pounds of refrigerant contained in the system, and the field test pressure applied.

Systems containing more than 100 lb of refrigerant should be provided with metal signs having letters of not less than ½ in. in height designating the main shut-off valves to each vessel, main steam or electrical control, remote control switch, and pressure limiting device. On all exposed high pressure and low pressure piping in each room where installed outside the machinery room, shall be signs as above with the name of the refrigerant and the letter HP or LP.

Marking.—Each separately sold refrigerant containing vessel larger than 5 cu ft in gross volume, each refrigerant condensing unit, and each refrigerant compressor shall carry a nameplate marked with the manufacturer's name and address, identification number, and name of refrigerant used.

Helmets.—One mask or helmet shall be required where amount of Group 2 refrigerants between 100 and 1,000 lb, inclusive, are employed. If more than 1,000 lb of Group 2 refrigerants are employed, at least 2 masks or helmets shall be required. Only complete helmets or masks marked as approved by the Bureau of Mines of the US Department of the Interior and suitable for the refrigerant employed shall be used and they shall be kept in a suitable cabinet immediately outside the machinery room or other approved accessible location.

Canisters or cartridges of helmets or masks shall be renewed immediately after having been used or the seal broken and if unused, must be renewed at least once every 2 yr. The date of filling shall be marked thereon.

Refrigerant Storage.—Not more than 300 lb of refrigerant in approved containers shall be stored in a machinery room.

No refrigerant shall be stored in a room in which less than 20 lb are used in the system.

Refrigerants on the user's premises in excess of that permitted in the machinery shall be stored in a fireproof shed or room used for no other purpose.

Charging and Discharging Refrigerants.—When refrigerant is added to a system, except a Unit System containing not more than 6 lb of refrigerant, it shall be charged into the low pressure side of the system. No container shall be left connected to a system except while charging or withdrawing refrigerant.

Refrigerants withdrawn from Refrigerating Systems shall only be transferred to approved containers. No refrigerant shall be discharged to a sewer. The containers from which refrigerants are discharged into or withdrawn from a Refrigerating System must be carefully weighed each time they are used for this purpose, and the containers must not be filled in excess of the permissible filling weight for such containers and such refrigerants as are prescribed in the Interstate Commerce Commission's "Regulations for the Transportation by Rail of Explosives and other Dangerous Articles in Freight, Express and Baggage Services including Specifications for Shipping Containers," effective October 1, 1930, with supplements 1 to 18 inclusive.

SAFETY INSTRUCTIONS FOR ICE AND COLD STORAGE PLANT OPERATION

Plant Operation

(a) Combustible materials should not be kept in the plant and in every case should be kept well away from liquid ammonia receivers.

(b) Good ventilators capable of rapid air changes should be in compressor room.

(c) Apply belt dressing only to that part of belt leaving a pulley. Liquid belt dressing applied with a long handle brush is much safer than stick belt dressing.

(d) Water or oil leakage from stuffing boxes should not be allowed to accumulate around equipment.

(e) All electrical equipment shall be grounded.

(f) Switchboards

    (1) All switchboards should be of marble or slate, or other dead front construction.

    (2) Current-carrying parts shall be enclosed with insulating material or grounded metal casings, so arranged to prevent accidental contact and to prevent short circuits.

    (3) A rubber mat or insulated platform shall be placed on floor in front of switchboard and extending its entire length.

    (4) No energized conductors or busses shall be touched or handled except by an authorized employee who wears properly tested rubber gloves.

    (5) Entrances to rear of switchboard shall be connected with gates. Warning signs forbidding entrance of unauthorized persons shall be placed on gates.

    (6) Meters or operating devices requiring attention under ordinary operating conditions every 24 hr or less, must be located in the above enclosures.

(g) More than one exit is necessary for quick escape in case of emergency.

(h) A master disconnect switch should be located in such manner that it can be pulled without entering building. When this switch is pulled it should stop every piece of apparatus in the plant in emergencies.

(i) Oil, gasoline, or other gaseous liquids, should not be used for cleaning purposes. Use carbon tetrachloride, Stoddard Solvent, or other special cleaning fluid. Do not use carbon tetrachloride in close or unventilated places.

(j) Scaffolds should be constructed to carry a center load of three times the maximum weight of men and material expected to be placed on it.

(k) If possible, eliminate necessity of employees crossing street to read meters.

Fire Prevention

    (1) Store all inflammable liquids in either an approved oil house, in a fire-proof room, or in fire-proof cabinets.

(2) All tanks, containers and cans shall be marked in some distinguishing manner, and are to be labeled as to contents.

(3) Never permit anyone to smoke, use open flames, or strike sparks where inflammable liquids are used or stored.

(4) Keep all oily rags and waste in approved metal containers.

(l) When natural gas leaks are suspected, apply a solution of soap and water to the exterior of the pipes and fittings with a brush. The soap suds will form bubbles at the point of the leak. Keep all flames, sparks, or lighted cigarettes, cigars and pipes away.

(m) On electric hand tools, an extra wire to ground the tool casing should be used. Use only approved and inspected extension cord.

(n) Air compressor tanks should be drained once a week during heavy season.

Guards

(a) The following machines, or machine parts shall be adequately guarded (see your insurance carrier for proper guards and/or State Standards):

(1) Belts, chains, rope drives.

(2) Friction drives, gears, sprockets.

(3) Vertical, inclined, or horizontal shafting.

(4) Fly wheels, clutches, couplings, projecting shaft ends.

(5) Emery wheels, motors or generators with exposed terminals, reciprocating machines, ice scoring machines, ice crushing machines.

Repairs or Installations

(a) Whenever it becomes necessary to work on any engine, pump generator, motor, balting, shafting, or other piece of machinery, a warning card—DANGER—MEN WORKING—must be properly attached to the governor, valve, throttle, switch, or other device used to set the machinery in motion.

(b) Machinery shall not be worked on or repaired while in motion.

(c) When working on or in crankcase of compressors make sure that employees, tools, extension cords or other impediments are clear before turning fly wheel.

(d) Employees moving heavy equipment on rollers should not wear gloves and rollers should be of sufficient length to extend well past the skids. (A glove finger may be caught under roller and hand pulled under.)

Tools

(a) Select the right tool for the job—never use a makeshift.

(b) Supervisors shall not permit the use of improper or unsafe tools or devices.

(c) All chisels, bars, drills, etc., which are held by one man and struck with a sledge hammer by another shall be held with long-handled tongs, a piece of garden, steam hose or other suitable device to prevent a glancing blow or a missed blow striking the employee holding the tool. The device used to hold the tool or bar should be at least 18 in.

long, and it should be held near the outer end. The holder shall wear goggles, and also, if possible, he should protect his face.

(d)   Tools should be kept on tool boards, tool racks or in tool boxes when not in use.

(e)   Select wrenches of the right size for the job. Face the jaws of an adjustable wrench in the direction of pull.

(f)   All edged or pointed tools shall be kept sharp. Dull tools are more dangerous than sharp ones.

(g)   Avoid using a hammer with hardened face on a highly tempered tool such as a drill, file, die, jig, etc. Chips may fly.

(h)   Never use a wrench on moving machinery; stop the machine.

(i)   Never use an extension on a wrench. The proper length of a wrench handle is calculated by its manufacturer for the strength of the wrench jaws.

Ladders

(a)   Employees shall not be permitted to use a ladder that has cracked or broken rungs, side pieces or straps.

(b)   Permanent ladders on towers, tanks, or where subject to water shall be of metal, or rungs shall be strapped to rails with metal plates. The plates or straps should be in one continuous piece.

(c)   Straight ladders shall be of sufficient length that work may be reached when standing on third or fourth rung from top.

(d)   Step ladders over 5 ft in height shall be of sufficient length that work may be reached when standing on second or third step from the top.

(e)   Distance from the foot of the ladder to the wall shall not exceed one-fourth of the distance from the top of ladder to the floor.

(f)   Ladders greater than 5 ft in length shall be blocked, anchored, or held by an assistant. All ladders should have nonslip shoes.

(g)   Improvised ladders such as barrels, boxes, chairs, tables, etc., shall not be used.

(h)   All stairways leading to the top of treating tanks, ammonia condensers, cooling towers or platforms shall be properly railed.

SAFETY DEVICES AND EQUIPMENT

(a) Gas Masks, Use and Availability

(1)   Every plant employee shall be instructed in the use of the ammonia gas mask. He should work in the mask periodically to become accustomed to its use.

(2)   One extra canister for the ammonia gas mask must be kept on hand at all times. Space canisters last for an indefinite period if the cork and seal have not been removed. Replace any canister which has become rusted.

(3)   In addition to using an ammonia gas mask when shutting off valves or otherwise stopping ammonia leaks, mask must be worn by operator when making any opening in the ammonia system. Masks shall be kept in cases

in a cool and convenient location. Place as near as possible to an outlet of a room. Upon attaching ammonia canister to gas masks, the seat at the bottom must be removed. Canisters should be replaced every 2 yr whether mask has been used or not.

(4) When using an ammonia mask, if the odor of the gas become objectionable the employee should immediately leave the gas area. He should replace the canister and examine the mask for defect before reentering.

(5) Mask should be placed in kit so that the fabric is not creased, the tube kinked, or valve bent.

(b) Goggles

(1) Employees using emery wheels or portable grinders shall wear goggles equipped with shatter-proof glass. They should be used at all times when chipping or cutting concrete, breaking up junk, porcelain, handling, demolishing or cleaning brickwork, and on all classes of material or fluids affording possibilities of injury to the eyes.

(2) Goggles of suitably colored glass shall be used when working with an acetylene torch or electric welder, and in any other place where eyes are exposed to excessive heat or glare.

(c) Rubber Gloves

Rubber gloves with gauntlet protectors must be kept in an easily accessible place and used if necessary to pull cut-outs or do other electrical work.

SAFETY DEVICES AND EQUIPMENT

Suitable Clothing

(a) Loose clothing should not be worn around moving machinery. Jumpers should be kept within overalls; keep sleeves down. Loose sleeves or trouser legs are also a hazard around machinery.

(b) Do not wear inflammable articles such as celluloid cap visors, celluloid eye glass frames, oily clothing, etc.

(c) Gloves shall not be worn around moving machinery.

(d) Avoid wearing metal articles, such as rings, watch or key chains, or metal cap visors.

(e) Keep good soles on shoes at all times. Shoes with thin soles will cause bruises and blisters and will allow the penetration of nails.

(f) Employees should wear safety shoes with safety toe caps. When working on icy or slippery surfaces, shoes with a special nonslip sole should be worn.

Other American Standards Association safety practice bulletins that are of special interest to refrigeration engineers are listed below. These are available at the office of the American Standards Association, 10 East 40th Street, New York City.

Short Title	A.S.A. Number
Abrasive Wheels	B7
Blower and Exhaust Systems	Z33.1
Building Exits Code	A9.1
National Electrical Safety Code (B.F.U. 70)	C1
Elevators, Dumbwaiters and Escalators	A17.1
Refrigeration Using Gas Fuel	Z21.19
Identification of Piping Systems, Scheme of	A13
Ladders, Safety Code	A14

"Safe Practice" pamphlets are available from National Safety Council, 20 North Wacker Drive, Chicago, Ill., on Safe Operation of Steam Boilers, Compressed Air Machinery, Cranes, Overhead Traveling, Goggles, and Power Transmission.

SAFETY IN DESIGN

The designer of each machine should incorporate safeguarding devices right into the machine. Too many machines must be provided with safety devices after they come from the manufacturer. It is very difficult to make satisfactory safety devices for a machine after it has been built.

Designers should visualize their design to see whether there is any part of their handiwork wherein an operator or repairman can be maimed for life by the crushing of a hand or finger in the moving parts which he must handle or repair frequently while the machine is in motion.

The temperature of bearings must be determined frequently, packing must be taken up and belts aligned. Provision should be made to do this with a minimum of hazard.

Projections from a machine may trip the operator. Ample aisle space should be provided so that there is no crowding of the machine operators into the danger zones.

Piping should be so designed that the expansion of the mains causes no undue strain at the compressor, condensers or evaporators. Very serious accidents have been caused by the rupture of a line or a valve at a point where the expansion pressure of the pipe has not been relieved by suitable provision of some kind.

The National Industrial Conference Board have issued "First Aid Suggestions" to their subscribers. Some of those of special interest to the refrigeration trade follow.

Treatment of Burns from Scalds, Fire, Electricity

Do not open blisters. Apply burn ointment, 3% bicarbonate of soda (baking soda) in vaseline, or cover burned area with sterile gauze wet with warm saline solution (teaspoonful of salt to 1 pint of water). Apply several thicknesses of clean gauze and bandage lightly. Send for physician.

Treatment of Acid Burns

Thoroughly flush all parts with water. Place under bath if necessary. Remove acid-soaked clothes. After acid has been thoroughly washed off, dry and apply burn ointment, 3% bicarbonate of soda in vaseline. Cover burned area with plenty of sterile gauze and bandage lightly. Send for physician.

Treatment of Alkaline Burns (such as those from lime, plaster, potash, ammonia, etc.)

Thoroughly flush all parts with water to remove all further damage from alkali. Place under a shower bath if necessary. After alkali has been thoroughly washed off, flood the part with a mixture of equal parts of vinegar and water and apply gauze dressing; bandage lightly and send for physician.

Treatment of Eye Injuries

No attempt should be made to remove a foreign body stuck in eye. In case of foreign body in, or injury to, the eye, apply a compress gauze dressing and bandage. In acid burns, wash freely with water and follow with 3% solution of bicarbonate of soda. In alkaline burns, lime, plaster, potash, ammonia, etc., flush with a mixture of granulated sugar and water. Send to physician at once.

Treatment of Unconscious Patients (any cause)

Place in comfortable position, flat on back, face to one side, keep warm. Watch for vomiting. If he vomits, be careful no foreign material enters the lungs. Loosen all tight clothing. Do not give anything to drink. Notify the physician.

Treatment for Suffocation, Electric Shock, Drowning, Gas Poisoning

Apply prone-pressure or Schafer method of artificial respiration.

SELECTED A.S.A. SAFETY INSTRUCTIONS ON INSTALLATION OF REFRIGERATION EQUIPMENT

Foundations and Supports for condensing units or compressor units shall be of substantial and noncombustible construction when more than 6 in. high.

Moving Machinery should be guarded in accordance with accepted safety standards.

Clear Space adequate for inspection and servicing of condensing units or compressor units shall be provided.

Condenser Units or Compressor Units with Enclosures shall be readily accessible for servicing and inspection.

Water Supply and Discharge Connections should be made in accordance with accepted safety and health standards.

Discharge water lines shall not be directly connected to the waste or sewer system. The waste or discharge from such equipment shall be over and above a trapped and vented plumbing fixture.

Illumination adequate for inspection and servicing of condensing units or compressor units should be provided.

Electrical Equipment and Wiring shall be installed in accordance with accepted safety standards.

TABLE 24.4

MAXIMUM PERMISSIBLE QUANTITIES OF FLAMMABLE REFRIGERANTS

Name	Chemical Formula	Maximum Quantity in Lb per Lb per 1,000 Cu Ft of Room Volume
Butane	C_4H_{10}	2.5
Ethane	C_2H_6	2.5
Ethyl chloride	C_2H_6Cl	6
Ethylene	C_2H_4	2
Isobutane	$(CH_3)_3CH$	2.5
Methyl chloride	CH_3Cl	10
Methyl formate	$HCOOCH_3$	7
Propane	C_3H_8	2.5

For refrigerants not listed in Table 24.4 the test pressure for the high pressure side shall be not less than the saturated vapor pressure of the refrigerant at 150°F. The test pressure for the low pressure side shall be not less than the saturated vapor pressure of the refrigerant at 110°F. However, the test pressure for either the high or low side need not exceed 125% of the critical pressure of the refrigerant. In no case shall the pressure be less than 30 psig.

Gas Fuel Devices and Equipment used with refrigerating systems shall be installed in accordance with accepted safety standards.

Machinery Room Requirements.—Each refrigerating machinery room shall be provided with tight-fitting door or doors and have no partitions or openings that will permit the passage of escaping refrigerant to other parts of the building.

Each refrigerating machinery room shall be provided with means for ventilation to the outer air. The ventilation shall consist of windows or door opening to the outer air, or of mechanical means capable of removing the air from the room. The amount of ventilation for refrigerant removal purposes shall be determined by the refrigerant content of the largest system in the machinery room.

A.S.A. SAFETY CODE INSTRUCTIONS ON DESIGN AND CONSTRUCTION OF EQUIPMENT

Aluminum, Zinc, or Magnesium shall not be used in contact with methyl chloride in a refrigerating system. Magnesium alloys shall not be used in contact with any halocarbon refrigerant.

Pressure Limiting Devices shall be provided on all systems containing more than 20 lb of refrigerant and operating above atmospheric pressures, and on all water cooled systems so constructed that the compressor or generator is capable of producing a pressure in excess of the test pressure; except water cooled unit systems containing not more than 3 lb of a Group 1 refrigerant providing the operating pressure developed in the system with the water supply shut off does not exceed $\frac{1}{5}$ the ultimate strength of the system, or providing an overload device will stop the action of the compressors before the pressure exceeds $\frac{1}{5}$ ultimate strength of the system.

TABLE 24.5

MINIMUM REFRIGERANT LEAK FIELD TEST PRESSURES

Refrigerant Name and Number[1]	Chemical Formula	Minimum Field Refrigerant Leak Test Pressures, Psig	
		High Side	Low Side
Ammonia (717)	NH_3	300	150
Butane (600)	C_4H_{10}	95	50
Carbon dioxide (744)	CO_2	1500	1000
Dichlorodifluoromethane (12)	CCl_2F_2	235	140
Dichlorodifluoromethane 73.8%) (500)	CCl_2F_2)	285	150
Ethylidene fluoride 26.2%)	CH_3CHF_2)		
Dichloroethylene (1130)	$C_2H_2Cl_2$	30	30
Dichloromethane (Methylene chloride) (30)	CH_2Cl_2	30	30
Dichloromonofluoromethane (21)	$CHCl_2F$	70	40
Dichlorotetrafluoroethane (114)	$C_2Cl_2F_4$	50	50
Ethane (170)	C_2H_6	1200	700
Ethyl chloride (160)	C_2H_5Cl	60	50
Ethylene (1150)	C_2H_4	1600	1200
Isobutane (601)	$[(CH_3)_3CH$	130	70
Methyl chloride (40)	CH_3Cl	210	120
Methyl formate (611)	$HCOOCH_3$	50	50
Monochlorodifluoromethane (22)	$CHClF_2$	300	150
Monochlorotrifluoromethane (13)	$CClF_3$	685[2]	685[2]
Propane (290)	C_3H_8	300	150
Sulfur dioxide (764)	SO_2	170	85
Trichloromonofluoromethane (11)	CCl_3F	30	30
Trichlorotrifluoroethane (113)	$C_2Cl_3F_3$	30	30

[1] ASRE Designation.
[2] Critical pressure is 561 psia at critical temp of 83.9°F.

BIBLIOGRAPHY

AMERICAN STANDARD ASSOC. 1941. Selection of accident factors for compiling industrial accident causes, ASA Z16.2.

CUTTER, W. A. 1951. Organization and functions of the safety department Res. Rept. 4. American Management Assoc.

GRIMALDI, J. V. 1951. Reducing costs through accident prevention engineering. Mech. Eng. 73, No. 6, 492–495.

HARRINGTON, D. 1935. Accident costs and safety dividends. US Bur. Mines Inform. Circ.

McFARLAND, R. A. 1953. Human Factors in Air Transportation. McGraw-Hill Book Co., New York.

MEIGS, R. R. 1958. Prevention of Accidents. Marks Mechanical Engineering Handbook, 6th Edition, 1259–1264.

NATIONAL SAFETY COUNCIL. 1951. Accident Prevention Manual. Chicago, Ill.

SIMONDS, R. H., and GRIMALDI, J. V. 1956. Safety Management: Accident and Cost Control. Richard D. Irwin, Homewood, Ill.

WOOLRICH, W. R. 1966. Accident prevention and safety in refrigeration. In Handbook of Refrigerating Engineering, 4th Edition, Vol. 2. Avi Publishing Co., Westport, Conn.

Appendix 1. Definitions of Terms Commonly Used in Refrigeration

INTRODUCTION

The managers, operators and maintenance men of cold and freezer warehousing are exposed to many terms especially related to the cold and freezer storage industry. Preparatory to presenting this manual to the men and women of the industry, the definitions common to the business, together with the conversion factors for international exchange of perishable foods are presented as an Appendix to this Manual.

COLD AND FREEZER STORAGE WAREHOUSE TERMINOLOGY AND DEFINITIONS

Absorbent.—An absorbing material which, due to its affinity for certain substances, extracts one or more such substances from a liquid or gaseous medium with which it is in contact and which changes physically or chemically, or both, during the process.

Absorber.—Usually a vessel used in an absorption refrigeration system to contain liquid for absorbing refrigerant vapors.

Absorption.—A process whereby a material extracts one or more substances present in an atmosphere or mixture of gases or other fluids.

Accumulator.—A storage tank for low-side liquid refrigerant, also known as a surge drum or surge header. Used in refrigerant circuit to reduce pulsation; to prevent liquid slop over to the compressor and as a positive liquid head on pumps used for liquid refrigerant recirculation. Most frequently found in ammonia systems although used, in some instances, in other refrigerant systems.

Activated alumina.—Aluminum oxide which absorbs moisture readily. Used as a drying agent.

Activated carbon.—A special form of porous carbon capable of absorbing various odors, anesthetics and similar vapors. Used in cold storage applications to absorb various room odors.

Adiabatic process.—A thermodynamic process during which no heat is extracted from or added to the thermal system.

Adsorbent.—A material which has the ability to cause molecules of gases, liquids or solids to adhere to its internal surfaces without changing the adsorbent physically or chemically. Certain solid materials such as silica gel are commercial adsorbents.

Adsorption.—The adherence of molecules of dissolved substances, gases or liquids to the adsorbent surfaces with which they are in contact.

Agitator.—A mechanical device to cause turbulence in a fluid within a tank. Usually in the form of a propeller, motor driven through a stuffing box. Can also take form of a pump placed outside the tank with suction and discharge piped to the tank to promote a liquid velocity inside the tank.

Air, ambient.—The air surrounding an object such as room air.

Air, dry.—Air with zero water vapor.

Air, saturated.—Air in which the partial pressure of water vapor is equal to the vapor pressure of water at the existing temperature, a condition of 100% relative humidity. A condition in which it is impossible to evaporate any additional water into the surrounding air.

Air, standard.—Air that weighs 0.075 lb per cu ft at 68°F dry bulb and 50% RH at a barometric pressure of 29.92 in. of mercury.

Air, cleaner.—A mechanical device used to remove airborne impurities by filtering. Used in heating and cooling systems.

Air diffuser.—A mechanical device designed to mix conditioned air from a duct with ambient air to obtain a uniform air condition. Usually also equipped with directional devices, either fixed or movable, to give a definite pattern to the resultant air stream.

Air washer.—A water-spray system for cleaning, humidifying or dehumidifying the air.

Algae.—A minute fresh water plant growth which forms a scum on the surface of recirculated water apparatus, interfering with fluid flow and heat transfer.

Analyzer.—A mechanical drum-like device connected in the high side of an absorption system for increasing the concentration of refrigerant in the vapor entering the rectifier or condenser.

Anemometer.—An instrument for measuring the velocity of air or other gas.

Atomize.—To reduce a liquid to a fine spray.

Baffle.—A partition, or partial partition, placed in an air stream or in any fluid flow to change or direct the direction of flow of the substance.

Barometer.—An instrument for measuring atmospheric pressure.

Bimetallic element.—An element formed by joining two metals having different coefficients of thermal expansion. Changes in temperature will cause the element to bend one way or the other. This motion can be transmitted mechanically to an electrical switch to form a thermostat.

Blanch.—To scald or sterilize (in the case of vegetables to be canned, cooked or frozen) usually by steaming or dipping in a hot-water bath at 212°F.

Bleeder pipe.—A tube bleeding a condenser or other drum or tank.

Boiler.—A closed vessel in which steam or other vapor is generated.

Boiling point.—The temperature at which a liquid vaporizes rapidly.

Booster.—A refrigerant compressor for low pressures, usually discharging, through an intercooler or desuperheater, into the suction of a higher pressure compressor. A means of two-stage refrigeration to gain efficiency in low temperature refrigeration.

Bore.—The inside diameter of a cylinder or tube.

Brazed.—The joining of two metals by using a soft or hard solder. Most often thought of in connection with high temperature solders such as silver bearing solder and bronze.

Brine.—A secondary refrigerant used for the transmission of heat without a change in its state. Usually in the form of a pumpable liquid circulated through some type of heat transfer surface in the cold area to extract heat from the cold area. The liquid brine can be a mixture of any substances with a freezing point below the temperature at which it is being used.

Brine cooler.—An evaporator for cooling brine in an indirect refrigeration system.

Brine cooler net refrigerating effect.—The product of the weight rate of brine flow and the difference in enthalpy of the entering and leaving brine expressed in heat units per unit of time. It is expressed also by the total refrigerating effect less the heat leakage losses.

Brine cooler total refrigerating effect.—The product of the weight rate of the refrigerant flow and the difference in enthalpy of the entering and leaving refrigerant fluid, expressed in heat units per unit of time.

Brine expansion tank.—A vented reservoir in a closed circulating brine system for volume expansion due to temperature change.

Brine return tank.—A reservoir in an open circulating brine system for for storage of brine at the pump suction.

Brine spray system.—A system in which cooling is achieved by a flow of air through a spray of cold brine. Complete units may consist of cooling coils to cool the brine pumped over the cooling surface and through the sprays, a fan to force air through the brine spray and an eliminator section to prevent the passage of brine droplets with the leaving air.

Brine system, closed.—A refrigerating system in which the circulating brine is completely enclosed and shut off from the atmosphere except for a vented expansion tank, usually located at the highest point of the system.

Brine tank.—In an ice plant, the main freezing tank in which the cans are immersed while ice is being produced. In a brine circulating system; a storage or balance tank for brine.

British thermal unit (Btu).—The heat required to produce a temperature rise of 1°F in 1 lb of water. The mean Btu is $\frac{1}{180}$ of the energy required to heat 1 lb of water from 32° to 212°F.

Bypass.—A pipe or duct, usually controlled by a valve or damper, for conveying a fluid around an element of a system.

Cabinet.—A mechanical refrigerator less the refrigerating machine.

Calibration.—The process of dividing and numbering the scale of an instrument; also of correcting or determining the error of an existing scale, or of evaluating one quantity in terms of readings of another.

Calorie.—The heat required to raise the temperature of 1 gm of water 1°C, actually from 4° to 5°C.

Calorimeter.—A device for measuring heat quantities, or steam quality, or heat of combustion. Also to measure specific heat, vital heat and heat leakage.

Capacity, refrigerating system.—The cooling effect produced by the change in total enthalpy between the refrigerating liquid entering the expansion device and the vapor leaving the evaporator, generally measured in Btu per hour or in tons of refrigeration.

Capillary tube.—A relatively small bore metal tube used in refrigeration to restrict flow and reduce pressure between the high and low sides. Used as a substitute for a thermal expansion valve. Usually used on systems of relatively small capacity such as refrigerators and small air conditioners up to five tons.

Carbon dioxide ice.—Solid CO_2 or dry ice.

Centigrade.—A thermometric scale in which the freezing point of water is 0° and the boiling point 100° at normal atmospheric pressure (14.696 psi).

Cargo batten.—A protection member applied permanently to the interior of a refrigerated compartment to provide air space between the stored cargo and the sides of the compartment.

Change of state.—Change from one phase, solid, liquid, or gas to another.

Charge.—Amount of refrigerant in a system; (to put the refrigerant charge in a system).

Chill.—To apply refrigeration moderately, as to meats, without freezing.

Chilled cargo.—Cargo maintained at an assigned temperature above the freezing point.

Chilling rooms.—Rooms where animal carcasses are chilled rapidly after dressing, prior to being placed in holding rooms for storage.

Clearance.—The space in the end of a compressor cylinder not occupied by the piston at the end of the compression stroke.

Clearance pocket.—In a compressor, a space of controlled volume to give the effect of greater or less cylinder clearance. Used to vary the capacity of a compressor.

Coefficient of expansion.—The change in length per unit length or the change in volume per unit volume per degree change in temperature.

Coefficient of heat transfer, overall (thermal transmittance).—The rate of heat flow per unit area under steady conditions through a body for a unit temperature difference of the fluid on the two sides of the body.

Coefficient of performance.—The ratio of the refrigeration produced to the heat energy required to obtain the refrigeration; driving components of the system having their power requirements expressed in heat energy terms.

Coil, cooling.—Piping or tubing which transfers heat from air or other gas to a refrigerant or brine.

Cold storage.—Preserving perishables on a large scale by refrigeration, usually that above 32°F.

Compression.—In a compression system of refrigeration, a process by which the pressure of the refrigerant gas is increased, usually by a reciprocating, centrifugal or rotary type compressor.

Compression, dual.—A split suction, or two suction pressures, carried on a single compressor by means of the compressor valving.

Compression, multi-stage.—Compression of a refrigerant gas in two or more steps, usually with the discharge of the low stage compressor discharging through appropriate intercoolers to the suction of a higher stage compressor.

Compression, ratio of.—The ratio of absolute pressures after and before compression.

Compression, single stage.—The compression of refrigerant gas from evaporator pressure to condensing pressure in one step.

Compression, wet.—A system in which some liquid remains in the refrigerant gas to the suction of the compressor. In extreme cases, this can cause slugging of the compressor with valve breakage and other bad effects.

Compressor, centrifugal.—A nonpositive displacement compressor which depends for pressure rise, at least in part, on centrifugal effect.

Compressor, compound.—A compressor with more than one cylinder in which compression is accomplished by stages.

Compressor, double acting.—A compressor which has two compression strokes per revolution of the crankshaft per cylinder.

Compressor, horizontal.—A compressor with a horizontal cylinder, or in small sizes, with a horizontal crankshaft.

Compressor, open-type.—A compressor with a shaft or other moving part extending through a casing to be driven by an outside source of power, thus requiring a stuffing box, shaft seals or equivalent rubbing contact between a fixed and moving part.

Compressor, positive displacement.—A compressor in which increase in vapor is attained by changing the internal volume of the compression chamber.

Compressor, reciprocating.—A positive displacement compressor with a piston or pistons moving in a straight line but alternately in opposite directions.

Compressor, rotary.—A compressor in which compression is obtained by the rotation of an off center solid cylinder within a fixed cylinder. The solid cylinder is slotted with blades operating in the slots free to move in and out following the fixed cylinder walls. Gas trapped between the blades is forced out after being compressed.

Compressor—single acting.—A compressor having one compression stroke per revolution of the crank for each cylinder.

Condensate.—The liquid formed by condensation of a vapor.

Condensation.—The process of changing a vapor into a liquid by the extraction of heat.

Condenser (liquefier).—A vessel or arrangement of pipes or tubing in which the vaporized (and compressed) refrigerant is liquefied by the removal of heat.

Condenser, air cooled refrigerant.—A refrigerant condenser cooled by natural or forced circulation of atmospheric air.

Condenser, atmospheric refrigerant.—A condenser with the cooling surfaces open to the atmosphere and cooled by a film of water flowing over the cooling surfaces.

Condenser, closed shell and tube.—A condenser with closed water heads at each end, the heads being equipped with partition webs to direct the water, under pressure, through the tubes in various combinations as required to obtain the desired water velocity in the tubes.

Condenser, evaporative refrigerant.—A heat transfer pipe or tubing coil placed in a casing equipped with water sprays and circulating pump and with a forced air fan to provide air circulation through the tube bundle. Heat removal is accomplished mainly by the evaporation of the spray water cooling the water which, in turn, removes heat from the refrigerant vapor in the coil.

Condenser, open shell and tube.—A condenser in which water passes in a film over the inner surfaces of the tubes, which are open to the atmosphere.

Condenser, submerged.—Condenser piping submerged in a bath of condenser water.

Condensing unit.—Usually the complete high-side of a refrigeration system consisting of compressor, condenser, driving motor, receiver and controls all mounted on a common base so that the entire combination can be moved as a single unit.

Conduction, thermal.—The process of heat transfer through a material medium in which kinetic energy is transmitted by the particles of the material from particle to particle.

Conductivity, thermal.—The ability of a material to transmit heat from one point in the material to another.

Conductivity, thermal (k factor).—The time rate of heat flow through a unit area of a homogeneous material under steady conditions when a unit temperature gradient is maintained in the direction perpendicular to the area. In English units, its value is usually expressed in Btu per (hr) (sq ft) °F per in. of thickness).

Conservation of energy.—A law of thermal dynamics that states that energy can be neither created nor destroyed.

Control, high pressure.—A pressure device (usually an electric switch) actuated mechanically by fluctuations in head pressure in a refrigeration system. This device usually used as a safety switch to shut down the plant in case the head pressure rises beyond safe limits.

Control, low pressure.—A pressure device actuated by the suction pressure of the refrigeration system. It is usually used as a control to prevent the suction pressure from attaining too low a level. It is also used to actuate other control devices such as compressor unloaders.

Convection.—The transfer of heat by the movement of a fluid containing thermal energy.

Cooler, cold storage.—An insulated room maintained at a temperature above 30°F.

Cooler, sensible heat.—A form of cooling surface using water, brine or direct expansion refrigerant. It is always located on the leaving side of the dehydrator and frequently treating, in addition, a large volume of room air which is not circulated through the dehydrator for moisture reduction.

Cooler unit.—A manufactured assembly usually including a cooling coil, fan (direct connected or belted) and associated controls, for installation in the space to be cooled.

Core area.—The total plane area of the portion of the grill, face, or register bounded by a line tangent to the outer edges of the outer openings through which the air can pass.

Corkboard.—Cork granules, cleaned, compressed and bonded by heat to a weight of about 0.65 lb per board ft, used for thermal insulation.

Counterflow.—In the process of heat exchange between two fluids, opposite direction of flow, coldest portion of one meeting the coldest position of the other.

Critical point.—The state point of a substance at which liquid and vapor have identical properties.

Cryohydrate.—A frozen mixture of water and a salt; a brine mix in eutectic proportions to give the lowest freezing point.

Cutting room (locker plants).—Cold room in which animal carcasses are cut up in commercial size roasts, steaks, etc.

Cycle.—Complete course of operation of a refrigerant back to a starting point measured in thermodynamic terms.

Cycle, Carnot.—A sequence of reversible processes forming the reversible cycle of an ideal heat engine of maximum thermal efficiency. It consists of isothermal expansion, adiabatic expansion, isothermal compression and adiabatic compression to the initial state.

Cycle closed.—Any cycle in which the primary medium is always enclosed and repeats the same sequence of events.

Cycle, defrosting.—A break in the regular refrigeration cycle to permit defrosting of the refrigerant evaporator surface. This can be accomplished in coolers above freezing by shutting off the refrigerant system and allowing the accumulated frost to melt. In rooms below freezing, artificial means such as hot refrigerant gas, water or electric heat may be employed to speed the defrosting.

Cycle, refrigeration.—The complete thermodynamic process in a refrigeration plant to produce the desired refrigeration effect.

Cycle, reversible.—Theoretical thermodynamic cycle which can be completely reversed; e.g., Carnot cycle.

Dalton's law of partial pressures.—The sum of the individual pressures of the constituents equals the total pressure of the mixture.

Decibel.—A unit commonly used for expressing sound or noise intensities referred to an arbitrary reference level.

Defrosting.—The removal of accumulated ice from a cooling element, also the thawing of frozen foods.

Degree day.—A unit determining the amount of heating required over a given period of time; equal, for any one day, to the difference between an arbitrary base temperature of 65°F and the average outside air temperature in °F over that 24-hr period.

Dehumidification.—The condensation of water vapor from air by cooling below the dew point or removal of water vapor from the air by chemical or physical means.

Dehumidifier.—An air cooler or washer used for lowering the moisture content of the air passing through it. An absorption or adsorption device for removing moisture from the air.

Dehydration.—The removal of water vapor from air by the use of absorbing or adsorbing materials, or the removal of water from stored foods.

Density.—The mass or weight of a substance per unit of volume.

Desiccant.—Any absorbent or adsorbent, liquid or solid, that will remove water or water vapor from a material.

Desiccation.—Any process for evaporating water or removing water vapor from a material.

Dewpoint.—Temperature of condensation if moist air is cooled at constant pressure.

Diesel engine.—An internal combustion engine used to change thermal to mechanical energy and using the heat of compression to ignite the fuel.

Diffuser, air.—An air delivery device arranged to promote mixing of the (primary) air leaving it with (secondary) room air by aspirating action. Also to give a pattern to the air circulation.

Dip tank.—A tank located conveniently to an ice can dump and supplied with water in which the ice can is immersed to thaw the ice block loose from the can.

Displacement.—Volume swept by a piston per revolution: usually expressed in cu in., or cu ft.

Displacement, actual.—The actual volume, at compressor inlet conditions, of gas moved per unit of time.

Displacement, theoretical.—The total volume displaced by all the pistons of a compressor at a given rpm expressed in cubic feet, or cubic inches, per unit of time.

Distributor.—A device for dividing the flow of liquid between parallel paths in an evaporator, or in other types of heat transfer apparatus.

Drier.—A manufactured device containing a desiccant, placed in the refrigerant circuit: its primary purpose being to collect and hold within the desiccant, all water in the system in excess of the amount which can be tolerated in the circulating refrigerant.

Drip Water.—Appears on thawing frozen food; water melting from an evaporator; water dripping from a cooling surface.

Dry ice.—Solid carbon dioxide, CO_2 (proprietary term).

Duct.—A passageway made of sheet metal or other suitable material, not necessarily leak-tight, used for conveying air or other gas at low pressure.

Dunnage.—Strips of wood or other material used in storage rooms to provide air space between the floor and packages and between packages of products.

Effective work—The net mechanical energy required by, or the load imparted to, a piston of a compressor.

Efficiency, mechanical.—The ratio of the compression energy of a compressor to the energy of the work input.

Electrical degree.—The 360th part of the angle subtended at the axis of the machine by two consecutive field poles of like polarity. One mechanical degree is thus equal to as many electrical degrees as there are pairs of poles in the machine.

Electrical precipitator.—A device for removing dust from the air by means of electric charges induced on the dust particles.

Emissivity.—Characteristic of a surface for giving off heat by radiation; equal to the absorptivity.

Energy, available.—The portion of the total energy that can be converted to work in a perfect engine.

Energy, internal (intrinsic).—The sum of all the kinetic and potential energies contained in a substance due to the states of motion and separation of its several molecules, atoms and electrons. It includes sensible heat (vibration energy) and the part of the latent heat that is represented by the increase in energy during evaporation.

Enthalpy.—A thermodynamic property of a substance defined as the sum of its internal energy plus the quantity Pv/J, where P = pressure of the substance; V = its volume and J = the mechanical equivalent of heat.

Entropy.—The ratio of the heat added to a substance to the absolute temperature at which it is added.

Enzyme.—A complex organic substance which can catalyze a chemical reaction, e.g., pepsin, and aids in digestion.

Equalizer.—A piping arrangement on an enclosed compressor to equalize refrigerant gas pressure in the crankcase and suction. A device for dividing the liquid refrigerant between parallel low side coils. Also a piping arrangement to divide the lubricating oil between the crankcases of compressors operating in parallel or in tandem. Any piping arrangement to maintain equal pressure in two or more vessels.

Eutectic mixture (solution).—A mixture which melts or freezes at constant temperature and with constant composition. The melting point of the mixture is usually the lowest possible for mixtures of given substances.

Evaporation.—The change of state from a liquid to a vapor.

Evaporative cooling.—The adiabatic exchange of heat between air and a water spray or wetted surface. The water assumes the wet bulb temperature of the air, which remains constant during its traverse of the exchanger.

Exhaust opening.—Any opening through which air is removed from a space which is being heated or cooled or humidified or dehumidified or ventilated.

Expansion coil, direct.—An evaporator constructed of pipe or tubing usually with hairpin bends to keep the coil compact. The surface may be plain (prime surface) or finned.

Factor of safety.—The ratio of ultimate stress to the design working stress.

Fahrenheit.—A thermometric scale in which 32° denotes freezing and 212° the boiling point of water under normal pressure at sea level (14.696 psi).

Fan, centrifugal.—A fan rotor or wheel within a scroll type housing and including a driving mechanism and supports for either belt drive or direct connection.

Fan, propeller.—A propeller or disc-type wheel within a mounting ring and including driving mechanism supports for either belt drive or direct connection.

Fan, vane axial.—A disc-type wheel within a cylinder with a set of air guide vanes located either before or after the wheel and including driving mechanism supports for either belt drive or direct connection.

Filter.—A device to remove solid material from a fluid.

Fin.—An extended surface such as metal sheets attached to tubes used to extend the surface of a tube and increase the heat transfer.

First law of thermodynamics.—A law stating the principle of the conservation of energy, equating heat and mechanical energy and denying perpetual motion.

Flammable refrigerant.—Any refrigerant that will burn when mixed with air, such as ethyl chloride, methyl chloride and the hydrocarbons.

Flash chamber.—A separating tank placed between the expansion valve and the evaporator in a refrigeration system to by-pass any flash gas around the evaporator.

Flash gas.—The refrigerant gas resulting from the very rapid evaporation of refrigerant in a pressure reducing device.

Flash point.—The temperature of a combustible material, as oil, at which there is a sufficient vaporization to ignite the vapor, but not sufficient vaporization to support combustion of the material.

Fluid.—Any gas, liquid or vapor.

Fluid, primary.—The refrigerant of the primary system that can be found in a gaseous state as well as a liquid state in different parts of the refrigeration system.

Fluid, secondary.—Usually a liquid that does not change state under the conditions as a secondary refrigerant and also usually cooled by the primary fluid.

Foaming.—The formation of a foam or froth of oil-refrigerant due to the rapid boiling of the refrigerant dissolved in the oil when the pressure is suddenly reduced.

Fog.—Suspended liquid droplets generated by condensation from the gaseous to the liquid state; or by breaking up a liquid into a dispersed state, such as splashing, foaming, or atomizing.

Free area.—The total minium area of the openings in a grille, face, or register through which air or gas can pass.

Freeze-up.—The failure of a refrigerating unit to operate normally due to the formation of ice at the expansion device.

Freezer storage.—An insulated room kept below 30°F. Cold storage room generally kept between +10°F and −10°F to receive and hold frozen foods. Sharp freezers are kept 10° to 20°F colder and receive unfrozen merchandise and freeze the merchandise to be placed for holding in a freezer storage room.

Freezing point.—The temperature at which a given liquid substance will solidify or freeze upon removal of heat.

Frost back.—The flooding of liquid refrigerant from an evaporator into the suction line accompanied by frost formation on the suction line.

Gas, inert.—A gas that neither experiences nor causes chemical reaction nor undergoes a change of state in a system or process.

Gas, noncondensable.—A gas in a refrigeration system that does not condense at the temperature and partial pressure at which it exists in the condenser, and therefore imposes a higher head pressure on the system.

Gas constant.—The coefficient R in the perfect gas equation $pv = RT$.

Grille.—The lattice or grating as a delivery or intake opening usually for air passage. Also shaped to give air flow pattern and direction in delivery air grilles.

Halide torch.—A leak tester utilizing a flame over a copper reactor plate. Under the influence of a halogen gas, the flame turns from blue to green. The flame is usually fed by alcohol, propane or acetylene.

Head, dynamic or total.—In a flowing fluid, the sum of the static and velocity heads at the point of measurement.

Head, static.—The static pressure of a fluid expressed in terms of the height of a column of the fluid, or of some manometric fluid, which it will support.

Heat, latent.—The heat quantity associated with change of state.

Heat, sensible.—The heat associated with a change in temperature.

Heat, specific.—The ratio of the quantity of heat required to raise the temperature of a given mass of any substance 1° to the quantity required to raise the temperature of a standard substance (usually water at 59°F) 1°.

Heat exchanger.—A device specifically designed to transfer heat between two physically separated fluids.

Heat of fusion.—The latent heat involved in changing between the solid and the liquid state.

Heat of the liquid.—The enthalpy of a mass of liquid above an arbitrary zero.

Heat of vaporization.—The latent heat required to change a liquid to a gaseous state.

Heat pump.—A reversible refrigeration system in which an exchange may be made between evaporator and condenser to alternately cool and heat an area.

Heat of reaction.—The heat per unit mass or per mol.

Heat transmission.—Any time-rate of heat flow, usually refers to conduction, convection and radiation combined.

Hermetically sealed compressor.—A compressor unit in which the compressor and compressor motor are enclosed in the same housing with no external shaft seal, the compressor motor operating in the refrigerant atmosphere.

High side.—The parts of a refrigeration system under condenser pressure or higher.

Hold-over.—In an evaporator, a material used to store heat or cold in latent or sensible form; to be used at a later time when the prime source of heat or cold is not available.

Horsepower.—A unit of power in the foot-pound second system. Work done at the rate of 550 ft lb per sec, or 33,000 ft lb per min.

Hot gas defrosting.—The use of high pressure or condenser gas in the evaporator or low side to effect the removal of frost.

Humidifier.—A device to add moisture to the air.

Humidify.—To add water vapor to the atmosphere; to add water vapor or moisture to any material.

Humidistat.—A control instrument responsive to atmospheric relative humidity, or an hygrostat.

Humidity, absolute.—The weight of water vapor per unit volume, as grains of water per cu ft of air.

Humidity, relative.—The ratio of the actual partial pressure of the water vapor in a space to the pressure of pure water at saturation at this temperature.

Humidity, specific.—The weight of water vapor (steam) associated with 1 lb of dry air, also called the humidity ratio.

Hydrometer.—An instrument which, by the extent of its submergence, indicates the specific gravity of the liquid in which it floats.

Hygroscopic.—Absorptive of moisture, readily absorbing and retaining moisture.

Hygrostat.—An automatic control responsive to humidity.

Ice melting equivalent.—The amount of heat absorbed by 1 lb of ice at 32°F in liquefying to water at 32°F — 143.4 Btu. (Note: the figure 144 Btu per lb is used in most calculations).

Inch of water.—A unit of pressure equal to the pressure exerted by a column of liquid water 1 in. high at a temperature of 4°C or 39.2°F.

Infiltration.—Air flowing inward as through a wall or crack.

Insulation, fill.—Granulated fibrous, shredded or powdered material, prepared from vegetable, animal or mineral origin, prepared in bulk or batt form to supply a barrier to the flow of heat.

Insulation, sound.—The acoustical treatment of housings, ducts and systems for isolation of vibration, or to reduce transmission of noise.

Insulation, thermal.—A material having a relatively high resistance to heat flow, and used principally to retard the flow of heat.

Intercooling.—The removal of heat from compressed gas between compression stages.

Interstellar cooling (sky cooling).—Radiation, especially on clear nights, from an exposed surface to the cold upper air strata.

Irradiation.—The subjecting of foods, etc., to radiation of special wavelengths, such as the 0.2537 μ wavelength, which destroys certain bacteria. Also the quantity of radiant energy incident on a surface per unit time and unit area.

Isentropic.—A reversible adiabatic process; a change taking place at constant entropy.

Isothermal.—A change taking place at constant temperature.

Jacket water.—The water used for cooling the cylinder head and/or walls of a compressor.

Joule-Thomson effect.—The fall in temperature which occurs when gas is allowed to expand without doing external work.

Leak detector.—A device used to detect leaks in a refrigerant system.

Liquefaction.—The change of state from gaseous to a liquid. This term generally used instead of condensation when referring to substances ordinarily gaseous.

Liquid line.—The tube or pipe carrying the refrigerant liquid from the condenser or receiver to the pressure reducing device in a refrigerant system.

Liquor.—A refrigerating solution used in absorption refrigeration.

Liquor, strong—Liquor, weak.—Relative terms of solution strength in an absorption refrigeration system.

Load factor.—The ratio of actual mean load to a maximum load or maximum production capacity in a given period.

Manometer.—A U-tube or single tube and reservoir, used with a suitable fluid to measure pressure differences.

Mechanical equivalent of heat.—An energy conversion ratio of 778.16 ft lb = 1 Btu.

Metabolism.—The chemical changes in living cells by which energy is provided for vital processes.

Modulating (of a control).—An adjustment by increments and decrements.

Mole (mol).—The weight of a substance equal to the weight units used in its molecular weight, as a gram mole, pound mole, etc.

Oil Separator.—A device for separating oil and oil vapor from the refrigerant, usually installed in the compressor discharge line.

Packing plant.—An establishment engaged in the slaughtering, dressing and processing of animals. Also used in connection with the processing of vegetables, fruits and fish.

Peltier effect.—The evolution of absorption of heat, which takes place when a current is passed across the junction between two different metals.

Performance factor.—In a mechanical condensing unit, the ratio of the capacity of the unit to the energy input, expressed in Btu per watt-hr, or in tons refrigeration per kilowatt.

Perm.—The unit of permeance. A perm is equal to 1 grain per (sq ft) (hr) (inch of mercury vapor pressure difference).

Phase.—In thermodynamics, one of the states of matter such as solid, liquid or gaseous. When used with electrical circuits, the type of current supplied such as single or three phase.

Plenum.—A chamber, under pressure, for receiving air, or other gas, prior to delivery to a conditioned space or combustion system.

Potentiometer.—An instrument for measuring or comparing small electromotive forces.

Power.—The rate of doing work.

Precooler.—A device or container for partially cooling a fluid, usually in advance of the regular cooling cycle.

Pressure, absolute.—Pressure measured above an absolute vacuum.

Pressure, atmospheric.—The pressure due to the weight of the atmosphere; the pressure indicated by a barometer.

Pressure, back or suction.—The operating pressure in a refrigeration plant, measured in the suction line at the compressor inlet.

Pressure, critical.—The vapor pressure corresponding to the critical temperature.

Pressure drop.—Loss in pressure due to friction of a fluid in a pipeline measured between two points in line of flow.

Pressure, dynamic.—The sum of the static pressure and the velocity pressure at the point of measurement.

Pressure, gage.—The pressure above atmosphere. The reading given by the usual Bourdon tube-type pressure gages found in most refrigeration plants.

Pressure, head.—The pressure in a refrigerating plant measured in the discharge line at the compressor outlet.

Pressure, partial.—The portion of total gas pressure, of a mixture, attributable to one component.

Pressure, saturation.—The pressure at which vapor and liquid, or vapor and solid, can coexist in stable equilibrium.

Pressure, total.—The sum of the static pressure and the velocity pressure at the point of measurement.

Pressure, vapor.—The pressure exerted by a vapor.

Psychrometer.—An instrument used to ascertain wet and dry bulb temperatures. From their readings are obtained the relative humidity and dew point.

Pump down—(refrigeration system).—The operation by which the refrigerant in a charged system is pumped into the liquid receiver.

Purger.—A device for removing noncondensable gases from a refrigerant high-side.

Pyrometer.—An instrument for measuring temperature.

Radiation, thermal.—The passage of heat from one surface to another without warming the space between. The heat is passed by wave motion similarly to the transmission of light.

Ratio of compression.—The ratio of absolute pressures after and before compression.

Refrigerant.—The medium of heat transfer in a refrigerating system which picks up heat by evaporating at a relatively low temperature and pressure, and gives up heat on condensing at a higher temperature and pressure.

Refrigeration.—The process of extracting heat from a substance or space by any means.

Refrigerator, commercial.—A refrigerator larger than a household box classified as reach-in, walk-in, self-service, display case, etc.

Regain of moisture.—The amount of moisture absorbed by any material in percentage of weight of that material.

Register.—A grille supplied with a control damper.

Respiration of living fruits and vegetables.—The production of CO_2 and heat by the ripening and breathing of perishables in storage.

Reversible thermal cycle.—A theoretical thermodynamic cycle which can be completely reversed, e.g., Carnot cycle.

Room dry bulb.—The ambient air temperature in a room as measured by a thermometer.

Sabin.—A unit of sound absorption equal to the equivalent absorption of 1 sq ft of a surface of unit absorptivity.

Salinometer.—A hydrometer calibrated in terms of salt concentration.

Saponify.—To turn to soap, properly the decomposition of an ester into an alcohol and an acid.

Sensible heat ratio, air cooler.—The ratio of the sensible cooling effect to the total cooling effect.

Silica gel.—A drier material having the basic formula, SiO_2. A sand-like substance.

Spray pond.—An arrangement for lowering the temperature of water by evaporative cooling of the water in contact with outside air.

Subcooling.—The cooling of a refrigerant below condensing temperature, for a given pressure. Also cooling a liquid below its freezing point, where it can exist only in a state of unstable equilibrium.

Sublimation.—A change of state directly from solid to gas.

Suction line.—The tube or pipe which carries the refrigerant vapor from the evaporator to the compressor inlet.

Sun effect.—Solar energy transmitted through space through windows and building materials.

Superheated vapor or gas.—Vapor at a temperature which is higher than the saturation temperature (i.e., boiling point) at the existing pressure.

Temperature, absolute.—The temperature expressed in degrees above absolute zero.

Temperature, dew point.—The temperature at which condensation starts if moist air is cooled at constant pressure with no loss or gain of moisture during the cooling process.

Temperature, dry bulb.—The temperature of a gas or mixture of gases indicated by an accurate thermometer when there is no heat flow to or from the thermometer bulb.

Temperature, wet bulb.—The temperature at which liquid or solid water (ice), by evaporating into air, can bring the air to saturation adiabatically at the same temperature.

Thermostat.—An automatic control device responsive to temperature.

Throttling of a fluid.—An irreversible adiabatic process which consists of lowering pressure by expansion without work.

Ton (of refrigeration) American.—A rate of heat interchange of 12,000 Btu per hour; 200 Btu per min.

Ton day of refrigeration American.—The heat removed by a ton of refrigeration for a period of 24 hr; 288,000 Btu. Approximately equal to the latent heat of fusion of melting 1 ton (2,000 lb) of ice, from and at 32°F.

Transmittance, thermal "U" factor.—The time rate of heat flow per unit area under steady conditions through a body for a unit temperature difference of the fluid of the two sides of the body. In English units its value is usually expressed in Btu per (hr) (sq ft) (°F temperature difference between the fluid on one side of the body and on the other side of the body).

Turbulent flow.—Flow in which the fluid moves transversely as well as in the direction of the flow.

Unit, hermetically sealed.—A refrigeration unit having no exposed mechanical driving connection and containing no external shaft seal.

Valve, check.—A valve allowing (fluid) flow in one direction only.

Valve, expansion.—A valve for controlling the flow of refrigerant to the cooling element.

Valve, purge.—A device to allow fluid to flow out of a system, particularly noncondensable gas.

Valve, pressure reducing.—A valve which maintains a uniform pressure on its outlet side.

Valve, suction.—A compressor valve which allows refrigerant gas to enter the cylinder from the suction line and prevents return flow.

Vapor.—A gas, particularly one near equilibrium with the liquid phase of the substance.

Vapor barrier.—A material with high resistance to the passage of water vapor. Used in insulation applications to prevent the migration of water vapor into and through the insulation.

Velocity, critical.—The velocity above which fluid flow is turbulent.

Viscosity.—The property of semifluids, fluids and gases to resist an instantaneous change of shape or arrangement of parts.

Viscosity, absolute.—The force per unit area required to produce a unit relative velocity between two parallel areas of fluid a unit distance apart.

Viscosity, kinematic.—The ratio of absolute viscosity to density of a fluid.

Vital heat.—The heat generated by fruits and vegetables in storage, due to an oxidation ripening.

Volatile liquid.—A liquid which evaporates readily at atmospheric pressure and room temperature.

Volume, specific.—The volume of a substance per unit mass; the reciprocal of density.

Walk-in refrigerator.—A refrigerated cooler or freezer with large entry doors suitable for foot traffic.

Water forecooling tank.—A tank where inlet water is cooled prior to freezing.

Water vapor.—A term used commonly in air conditioning parlance to refer to steam or moisture in the atmosphere.

Weep.—The drip from frozen foods.

Wet bulb depression.—The difference between dry bulb and wet bulb temperature.

Wire drawing.—The restriction of the conductor for a flowing fluid, causing a loss in pressure by friction without loss of heat or performance of work; throttling.

Wort.—The unfermented malt that when fermented, produces beer.

BIBLIOGRAPHY

Definitions and Technical Terms Related to Cold and Freezer Warehouse Storage, Including Air Conditioning, Sanitation and Thermal Processing

AMERICAN SOCIETY OF HEATING, REFRIGERATING AND AIR CONDITIONING ENGINEERS. 1965–1966. ASHRAE Guide and Data Books, Fundamentals and Equipment. Chapter 67.

AMERICAN SOCIETY OF REFRIGERATING ENGINEERS. 1957–58. Air Conditioning Refrigerating Data Book, 10th Edition. Chapter 39, 1–27.

AMERICAN STANDARDS ASSOCIATION. 1958. Refrigeration Terms and Definitions, A.S.A. Bull. *B 53.1*. New York.

ANON. 1964. Thesaurus of Engineering Terms. Engineers Joint Council, New York.

TWENEY, C. F., and HUGHES, L. E. C. 1965. Chamber's Technical Dictionary, 3rd Edition. Macmillan Co., New York.

Appendix 2. Useful Conversion Factors

BRITISH TO METRIC AND METRIC TO BRITISH, ESPECIALLY RELATED
TO FREEZER STORAGE AND OPERATIONAL REFRIGERATION

Length	1 m = 39.37 in. = 3.2808 ft
Area	1 sq m = 10.7638 sq ft
Volume	1 cu ft = 1,728 cu in. = 28.32 liters
Density	1 gm per cu m = 0.3613 lb per cu in. = 62.4238 lb per cu ft = 8.3454 lb per gal. (US)
Mass and weight	1 kg = 15432.4 grains = 2.2046 avoirdupois lb
Pressure	1 lb per sq in. = 0.0703 kg per sq cm = 8.0703 m column of water = 2.3066 ft column of water
Velocity	1 mile per hr = 1.4666 ft per sec = 0.8684 knots = 1.6094 km per hr. Standard gravity = 32.17 (ft) (sec²) = 980.67 (cm) (sec²)
Energy	1 US hp hr = 1,980,000 ft lb = 273,745 kg m = 2,544.65 Btu or 641,240 kgm cal
Energy and heat	1 hp = 550 ft lb per sec = 76.0404 kg m per sec = 0.74565 kw.
Viscosity	1 gm per cm sec or dyne sec per sq cm poise = 100 lb sec per sq ft
Enthalpy	1 kg cal per kg = 1.8 Btu per lb. 1 Btu per lb = 0.5556 kg cal per kg.
Entropy	1 kg cal per (kg) °C) = 1 Btu per (lb) (°F).
Thermal conductivity	1 Btu per (hr) (ft) (°F) = 1.487 kg cal per (hr) (m) (°C)
Coeff. of heat trans	1 Btu per hr per sq ft °F = 4.88 kg cal per (hr) per (sq m) per °C = 1 Chu[1] per (hr) (sq ft) per °C

[1] The centigrade heat unit Chu is the heat required to raise 1 lb of water 1°C and equals 1.8 Btu.

CONVERSION TABLE

Engineering Gravitational Units
Units of Mass, Lb and Kg Units of Weight or Force, Lb$_f$ and Kg$_f$

	British	Metric Equiv.
Time	1(hr)	1(hr)
Length	1(in.)	25.4(mm)
	(ft)	0.3048(m)
	(mile)	1.610(km)
Surface	1(sq ft)	0.0929(m²)
Volume	1(cu ft)	0.02832(m³)
	(qt)	0.9464(l)
	(gal)(US)	3.785(l)
	(Imp)	4.536(l)
Mass	1(lb)	0.4536(kg)
Force or Weight	1(lb$_f$)	0.4536(kg$_f$)
Density	1 (lb)/(ft³)	16.018 (kg)/(m³)
	1 (lb)/(gal US)	0.1198 (gm)/(cm³)
Pressure	1 (lb$_f$)/(ft²)	4.883 (kg$_f$)/(m²)
	1 (lb$_f$)/(in.²)	703.1 (kg$_f$)/(m²)
Flow Rate	1 (ft³)/(min)	1.699 (m³)/(hr)
Dynamic Viscosity	1 (lb)/(hr) (ft)	1.488 (kg)/(hr)(m)
	1 (lb$_f$)(hr)/(ft²)	4.88 (kg$_f$)(hr)/(m²)
Kinematic Viscosity	1 (ft²)/(hr)	0.0929 (m²)/(hr)

TABLE A2.2—(*Continued*)

CONVERSION TABLE

Energy or Heat	1 (Btu)	0.2520 (kcal)
	1000(Btu)	0.293 (kw)(hr)
	1000(ft)/(lb$_f$)	0.3241 (kcal)
Power	1(hp)	1.014 metric hp
	1 (Btu)/(sec)	1.054 (kw)
Temperature	1 °F	0.5556 (°C)
	°F = 1.80(°C) + 32°	
	°R = 459.7 + (°F)	
Specific Heat	1 (Btu)/(lb)(°F)	1.0(kcal)/(kg)(°C)
Enthalpy, Latent Heat	1 (Btu)/(lb)	0.5556(kcal)/(kg)
Thermal Conductivity	1 (Btu)/(hr)(ft)(°F)	1.487 (kcal)/(hr)(m)(°C)
Thermal Diffusivity	1 (ft²)/(hr)	0.0929(m²)/(hr)
Heat Transf. Coeff.	1 (Btu)/(hr)(ft²)(°F)	4.88(kcal)/(hr)(m²)(°C)
Refrigeration	1 ton(US)	3024(kcal)/(hr)
	1 ton(Brit)	1.0(kcal)/(sec)

	Metric	British Equiv.
Time	1(hr)	1(hr)
Length	1 (cm)	0.3937 (in.)
	1 (m)	3.2808 (ft)
	1 (km)	0.6214 (mile)
Surface	1 (m²)	10.764 (ft²)
Volume	1 (m³)	35.314 (ft³)
	1 (l)	1.057 (qt)
	1 (l)	0.2642 (gal US)
	1 (l)	0.220 (gal Imp)
Mass	1 (kg)	2.2046 (lb)
Force or weight	1 (kg$_f$)	2.2046 (lb$_f$)
Density	1 (kg)/(m³)	0.06243 (lb)/(ft³)
	1 (gm)/(cm³)	8.35 (lb)/(gal US)
Pressure	1 (kg$_f$)/(m²)	0.2048 (lb$_f$)/(ft²)
	1 (kg$_f$)/(cm²)	14.22 (lb$_f$)/(in.²)
Flow rate	1 (m³)/(hr)	0.589 (ft³)/(min)
Dynamic viscosity	1 (kg)/(hr)(m)	0.672 (lb)/(hr)(ft)
	1 (kg$_f$)(hr)/(m²)	0.205 (lb$_f$)(hr)/(ft²)
Kinematic viscosity	1 (m²)/(hr)	10.764 (ft²)/(hr)
Energy or heat	1 (kcal)	3.968 (Btu)
	1 (kw)(hr)	3415 (Btu)
	1 (kcal)	3090(ft)(lb$_f$)
Power	1 metric hp	0.9863(hp)
	1 (kw)	0.9486 (Btu)/(sec)
Temperature	1 °C	1.800 (°F)
	°C = ⁵⁄₉(°F − 32°)	
	°K = 273.2 + (°C)	
Specific heat	1 (kcal)/(kg)(°C)	1.0(Btu)/(lb)(°F)
Enthalpy, latent heat	1 (kcal)/(kg)	1.80 (Btu)/(lb)
Thermal conductivity	1 (kcal)/(hr)(m)(°C)	0.672 (Btu)/(hr)(ft)(°F)
Thermal diffusivity	1 (m²)/(hr)	10.764 (ft²)/(hr)
Heat transf. Coeff.	1 (kcal)/(hr)(m²)(°C)	0.205 (Btu)/(hr)(ft²)(°F)
Refrigeration	1000(kcal)/(hr)	0.331 ton refrig (US)
	1 (kcal)/(sec)	1.0 ton refrig (Brit)

TABLE A2.2—(*Continued*)

CONVERSION TABLE

Standard gravity	32.17(ft)/(sec²)	980.67 (cm)/(sec²)
Standard atmosphere	2116.8 (lb$_f$)/(ft²)	10333(kg$_f$)/(m³)

1 Ton refrigeration	US 288,000 Btu/24 hr = 12,000 Btu/hr = 200 Btu/min
	US 0.840 ton refrig, Brit = 0.840 Frigorie
	Brit 342,800 Btu/24 hr = 14,285 Btu/hr = 238 Btu/min = 1 kcal/sec
	Brit 1.190 ton refrig, US = 1 Frigorie
British thermal unit	777.5 ft-lb$_f$
1 boiler horsepower	33,479 Btu/hr
Pound (force) per square inch	0.0684 atm = 2.036 in. Hg = 51.3 mm Hg
1 atmosphere	14.70 lb$_f$/in.² = 29.92 in. Hg = 760 mm Hg
1 Centipoise	2.419 lb/ft-hr = 3.60 kg/m-hr
1 Horsepower (US)	0.7457 kw = 42.44 Btu/min = 33.000 ft-lb$_f$/min = 550 ft-lb$_f$/sec
1 Metric horsepower	75 kg$_f$-m/sec = 1 Ps = 1 chev-vap
1 cubic foot	7.481 US gal = 28.315 l
1 US gallon	0.1337 cu ft = 0.835 Imp gal
1 Imperial gallon	0.1604 cu ft = 1.20 US gal

Source: W. B. Van Arsdel 1967.

TABLE A-2.3

RELATED TEMPERATURE CONVERSIONS

The Numbers Between Centigrade and Fahrenheit Columns Refer to the Temperature Either in Centigrade or Fahrenheit which is to be Converted to the Other Scale

C	C or F	F	C	C or F	F	C	C or F	F	C	C or F	F	C	C or F	F
−40.0	−40	−40.0	−23.3	−10	+14.0	−6.7	+20	+68.0	+10.0	+50	+122.0	+26.7	+80	+176.0
−39.4	−39	−38.2	−22.8	−9	+15.8	−6.1	+21	+69.8	+10.6	+51	+123.8	+27.2	+81	+177.8
−38.9	−38	−36.4	−22.2	−8	+17.6	−5.5	+22	+71.6	+11.1	+52	+125.6	+27.8	+82	+179.6
−38.3	−37	−34.6	−21.7	−7	+19.4	−5.0	+23	+73.4	+11.7	+53	+127.4	+28.3	+83	+181.4
−37.8	−36	−32.8	−21.1	−6	+21.2	−4.4	+24	+75.2	+12.2	+54	+129.2	+28.9	+84	+183.2
−37.2	−35	−31.0	−20.6	−5	+23.0	−3.9	+25	+77.0	+12.8	+55	+131.0	+29.4	+85	+185.0
−36.7	−34	−29.2	−20.0	−4	+24.8	−3.3	+26	+78.8	+13.3	+56	+132.8	+30.0	+86	+186.8
−36.1	−33	−27.4	−19.4	−3	+26.6	−2.8	+27	+80.6	+13.9	+57	+134.6	+30.6	+87	+188.6
−35.6	−32	−25.6	−18.9	−2	+28.4	−2.2	+28	+82.4	+14.4	+58	+136.4	+31.1	+88	+190.4
−35.0	−31	−23.8	−18.3	−1	+30.2	−1.7	+29	+84.2	+15.0	+59	+138.2	+31.7	+89	+192.2
−34.4	−30	−22.0	−17.8	0	+32.0	−1.1	+30	+86.0	+15.6	+60	+140.0	+32.2	+90	+194.0
−33.9	−29	−20.2	−17.2	+1	+33.8	−0.6	+31	+87.8	+16.1	+61	+141.8	+32.8	+91	+195.8
−33.3	−28	−18.4	−16.7	+2	+35.6	.0	+32	+89.6	+16.7	+62	+143.6	+33.3	+92	+197.6
−32.8	−27	−16.6	−16.1	+3	+37.4	+0.6	+33	+91.4	+17.2	+63	+145.4	+33.9	+93	+199.4
−32.2	−26	−14.8	−15.6	+4	+39.2	+1.1	+34	+93.2	+17.8	+64	+147.2	+34.4	+94	+201.2
−31.7	−25	−13.0	−15.0	+5	+41.0	+1.7	+35	+95.0	+18.3	+65	+149.0	+35.0	+95	+203.0
−31.1	−24	−11.2	−14.4	+6	+42.8	+2.2	+36	+96.8	+18.9	+66	+150.8	+35.6	+96	+204.8
−30.6	−23	−9.4	−13.9	+7	+44.6	+2.8	+37	+98.6	+19.4	+67	+152.6	+36.1	+97	+206.6
−30.0	−22	−7.6	−13.3	+8	+46.4	+3.3	+38	+100.4	+20.0	+68	+154.4	+36.7	+98	+208.4
−29.4	−21	−5.8	−12.8	+9	+48.2	+3.9	+39	+102.2	+20.6	+69	+156.2	+37.2	+99	+210.2
−28.9	−20	−4.0	−12.2	+10	+50.0	+4.4	+40	+104.0	+21.1	+70	+158.0	+37.8	+100	+212.0
−28.3	−19	−2.2	−11.7	+11	+51.8	+5.0	+41	+105.8	+21.7	+71	+159.8	+38.3	+101	+213.8
−27.8	−18	−0.4	−11.1	+12	+53.6	+5.5	+42	+107.6	+22.2	+72	+161.6	+38.9	+102	+215.6
−27.2	−17	+1.4	−10.6	+13	+55.4	+6.1	+43	+109.4	+22.8	+73	+163.4	+39.4	+103	+217.4
−26.7	−16	+3.2	−10.0	+14	+57.2	+6.7	+44	+111.2	+23.3	+74	+165.2	+40.0	+104	+219.2
−26.1	−15	+5.0	−9.4	+15	+59.0	+7.2	+45	+113.0	+23.9	+75	+167.0			
−25.6	−14	+6.8	−8.9	+16	+60.8	+7.8	+46	+114.8	+24.4	+76	+168.8			
−25.0	−13	+8.6	−8.3	+17	+62.6	+8.3	+47	+116.6	+25.0	+77	+170.6			
−24.4	−12	+10.4	−7.8	+18	+64.4	+8.9	+48	+118.4	+25.6	+78	+172.4			
−23.9	−11	+12.2	−7.2	+19	+66.2	+9.4	+49	+120.2	+26.1	+79	+174.2			